MENSUN BOUND

THE SHIP BENEATH THE ICE

The Discovery of
SHACKLETON'S
ENDURANCE

PAN BOOKS

First published 2022 by Macmillan

This edition first published 2023 by Pan Books
an imprint of Pan Macmillan
The Smithson, 6 Briset Street, London ECIM 5NR
EU representative: Macmillan Publishers Ireland Ltd, 1st Floor,
The Liffey Trust Centre, 117–126 Sheriff Street Upper,
Dublin 1, DOI YC43
Associated companies throughout the world
www.panmacmillan.com

ISBN 978-1-0350-0842-1

1 3 5 7 9 8 6 4 2

A CIP catalogue record for this book is available from the British Library.

Penguin illustration on p. *iii*, Sabertooth drawing on p. 214
and *Endurance* diagram on p. 330 by Hemesh Alles.
Map artwork on p. *ix* by ML Design Ltd.

Typeset by Palimpsest Book Production Ltd, Falkirk, Stirlingshire
Printed and bound by CPI Group (UK) Ltd, Croydon, CRO 4YY

Visit **www.panmacmillan.com** to read more about all our books
and to buy them. You will also find features, author interviews and
news of any author events, and you can sign up for e-newsletters
so that you're always first to hear about our new releases.

To my wife Jo
This one's for you

CONTENTS

Map of the *Endurance*'s journey ix

Author's note 1

Introduction 3

PART ONE: THE WEDDELL SEA EXPEDITION 2019

January 2019 9

February 2019 97

PART TWO: THE ENDURANCE22
EXPEDITION 2022

January 2022 207

February 2022 209

March 2022 293

Epilogue 373

Appendix 1: Members of the Imperial Trans-Antarctic
Expedition 1914–17 377

Appendix 2: Members of the Weddell Sea Expedition 2019 379

Appendix 3: Members of the Endurance22 Expedition 2022 383

Acknowledgements 387

Text credits 397

Picture credits 399

Selected Shackleton bibliography 401

Index 405

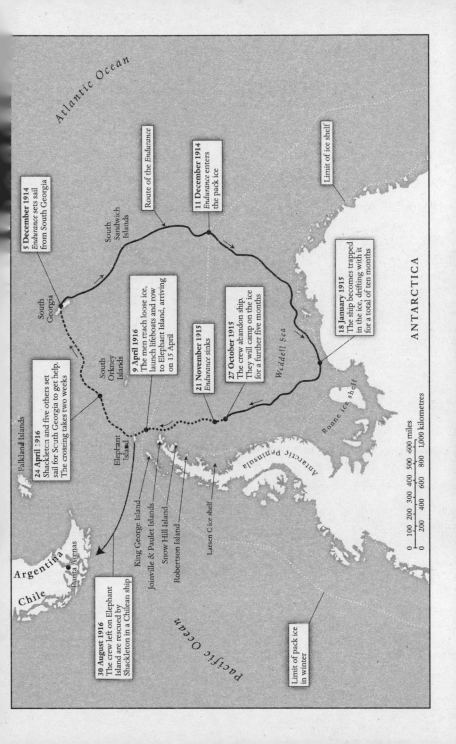

AUTHOR'S NOTE

The story that unfolds between these covers is *my* story. I stress the word 'my' because I know that others will remember it differently. Over our two campaigns, there were more than 150 people on the S.A. *Agulhas II*. Each will have his or her own story, and all of those stories will be as valid as my own.

Earlier this year, my brother Graham published his day-by-day diary of the 1982 Falklands War. He too struggled over how best to chronicle his perceptions of what happened; in the introduction, he wrote: 'No single memoir will be completely accurate, as authors write from different viewpoints with different biases.' Like my brother, I have had to wrestle with memory, perception and the fact that, in 24-hour operations over a great many days, much happened that I did not directly witness.

What follows is based upon my blogs, diary entries, daily reports, social media posts, announcements, emails, interviews and notebooks; but all that notwithstanding, there is a large subjective component. The search for the *Endurance* was highly complex, and reconciling everything that happened into a narrative that reflects everybody's experience would be impossible. Put another way, if the writers of the Gospels could not agree on detail, then what hope have I? So, again, I want to emphasize that the story here is told from my own perspective and the views I have expressed are mine alone. I hope it is clear that this was never intended to be – nor should it be seen as – an official record, and any errors or omissions are my responsibility.

It might be said that this book exists in part because of the Covid-19 pandemic. The first part was written in Port Stanley during lockdown in 2021, when I suddenly found myself housebound with nothing to do. That led on to the second part, which was written on the voyage back from Antarctica and at home in Oxford.

'And what about the *Endurance*?' These were the words of a friend and colleague in a Kensington coffee shop exactly ten years before I write this. That was the moment of inception. A long period of gestation followed, during which the necessary technical capability was assembled for our expedition; and then, on 6 April 2017, everything moved into preliminary planning and first-phase mobilization. The first campaign – the Weddell Sea Expedition – took place in 2019 under the Flotilla Foundation, and the next, Endurance22, in 2022 under the Falklands Maritime Heritage Trust (FMHT).

The 2019 campaign was mainly scientific, which accorded perfectly with the aims of the Flotilla Foundation; the objectives of the 2022 campaign were mainly archaeological and historical, reflecting the Trust deed of the FMHT. The two campaigns shared a common denominator that was fundamental to everything: the involvement of the marine robotics company Ocean Infinity, which provided not only the know-how, equipment and technicians, but also much of the inspiration.

Finally, special mention must be made of the teams (listed in the appendices) who conducted the search. Without complaint, they faced extremely aggressive ice conditions in 2019 and profound cold (−40°C) in 2022. They were mostly British, South African, French, American or German, with a sprinkling of other nationalities. Although the 2019 team were not there at the moment of discovery, they were with us in spirit. What we learned in the first campaign was essential for the success of the second. From beginning to end, it was a collective effort to which everybody contributed equally.

So let me end by saying that on our grand quest to find the *Endurance,* I feel privileged to have walked with champions. Champions all.

Mensun Bound
Director of Exploration
Oxford, August 2022

INTRODUCTION

On 5 December 1914, the 40-year-old British Antarctic explorer Sir Ernest Henry Shackleton set sail from the relative safety of South Georgia in the tempestuous Southern Atlantic on the ship *Endurance*, with a crew of 26 men, a stowaway, almost 70 dogs and a ship's cat.

Shackleton's expedition was rather grandly entitled the 'Imperial Trans-Antarctic Expedition', and his intention was to cross the Antarctic Continent from sea to sea via the South Pole. This journey had been meticulously planned and prepared for, with supply depots en route and another ship, the *Aurora*, to be waiting for the explorers on the other side of the continent. In an age of heroic expeditions, the world's media were gripped by the privations and harsh conditions the men might face, and the presence of photographer and filmmaker Frank Hurley on board meant that fame and celebrity beckoned for those taking part in this uniquely difficult challenge.

The inhospitable Weddell Sea, however, had other ideas. Just six weeks after the *Endurance* left South Georgia, the temperature plummeted and she became trapped in the sea ice many miles from the Antarctic Continent itself. The ship and the men on board were at the mercy of the weather and the frozen sea.

For many months, through the darkness and horror of an Antarctic winter, Shackleton's men lived on board, eking out an existence from the supplies on the ship. They hoped the following spring would bring respite, but instead, as the floes weakened around the ship, the movement of the ice put tremendous pressure on her hull and she began to splinter like a walnut in a vice. Shackleton gave the order to abandon ship. The crew rescued what supplies they could and on 21 November 1915, the *Endurance* finally sank beneath the ice.

For the rest of that Antarctic summer the men camped out on a large ice floe, hoping that it would drift them to safety and rescue, but by April 1916 the floe was breaking up and any sort of rescue seemed unlikely. With dwindling morale and supplies, Shackleton gave the order to take to the three lifeboats. Then, through a combination of luck and the remarkable navigational skills of the former captain of the *Endurance*, Frank Worsley, the crew made the five-day, 346-mile open sea journey to the uninhabited Elephant Island.

Their only hope of rescue lay back in South Georgia, from where the *Endurance* had set out, so Shackleton took one of the lifeboats and five crew members to make the treacherous 800-mile voyage there. Through towering seas, at times enduring hurricane-force winds, with Worsley plotting their position using the rudimentary technology of the day, the boat finally came in sight of South Georgia. Even then, their ordeal was not over. The seas were too rough to debark safely near human habitation, so they had to land on the unoccupied southern shore. The three strongest men then traversed the island's mountainous interior on foot, a crossing so dangerous that it would not be repeated until 1955. On 20 May they reached the whaling station at Stromness, where plans were put in place to rescue the three men stranded on the south shore and the 22 remaining on Elephant Island.

Any way you look at it, this is an extraordinary story of courage, determination and, of course, endurance. The iconic ship on which the men travelled was indeed aptly named. The fact that they did not complete what they originally set out to do has rightly faded from view. What is incredible is that, under ferociously hostile conditions, not a single crew member was lost. They had the strength and fortitude – both mental and physical – and the skills, organization and leadership to defy the odds.

The ship herself, the *Endurance*, became a symbol of that resistance and courage. During the harsh Antarctic winter of 1915, trapped in the ice though she was, the *Endurance* was a place of safety and refuge. When she finally went down, the pain and sorrow expressed in the diaries kept by various crew members is palpable. Frank Hurley's photographs of the ship trapped in the ice, her damaged masts pointing skyward like arthritic fingers, have become the lasting images of the expedition.

I grew up in the South Atlantic, on the Falkland Islands. Everyone in my generation was a Shackleton enthusiast. Shackleton had travelled to the Falklands three times and once tried to set up a sealing enterprise there. With Frank Worsley, he had even stayed at an inn once managed by my great-great-uncle, Vincent Biggs. The guestbook bearing their signatures still survives to this day. South Georgia, where Shackleton's most famous expedition began and ended, was our closest easterly neighbour at just under 1,000 miles away.

Shackleton has always loomed large in my life. My first introduction to his story was through my father, and later, when I was about seven or eight, I was given Webster Smith's book *Sir Ernest Shackleton* as a prize for Sunday school attendance. I still have it, and took it with me on the Endurance22 expedition.

As a fifth-generation Falkland Islander, I was born of the sea, so it's perhaps no surprise that I became a marine archaeologist specializing in shipwrecks and lost underwater worlds. I have dived on countless wrecks around the world. Each is different, each with its own unique character and story to tell about the men and women who travelled on board, about the times in which they lived, and about the moments before the ship went down.

But the elusive wreck of the *Endurance* has always held a special allure. I have always wondered what secrets she might hold, but for all the challenges I'd faced in my career, finding where she lay entombed beneath the ice of the Weddell Sea seemed an impossible dream – that is, until quite recently, when subsea robotics technology reached a level of development that meant it might just be capable of coping with everything the Weddell Sea could throw at us. A colleague and friend suggested we attempt the impossible, and a specialist team was slowly gathered.

As I pored over the diaries of Shackleton's men, I began to

believe that there was enough detail and evidence in their pages to help us pinpoint the exact whereabouts of the ship under the ice. The planning and preparation of that first expedition took many years, but on 1 January 2019, I finally found myself on board the modern icebreaker S.A. *Agulhas II*. Together with an incredible crew I was ready to embark on a journey to find the greatest shipwreck of them all: the *Endurance*.

PART ONE

The Weddell Sea Expedition 2019

1 JANUARY 2019

Preparations for departure

There were 27 of them and Shackleton.

In August 1914, as the First World War erupted, they set off on an expedition that would become the stuff of legend. Their goal was to cross Antarctica on foot, from the Weddell Sea to the Ross Sea by way of the Pole. It was, in Shackleton's words, 'the last great journey left to man'. But they never got started; in fact, they never even set foot upon the Great White Continent. In January 1915, while they were carving their way through the pack, their ship – the famed *Endurance* – became icebound, and eventually she was crushed. Ten months later, on 21 November, they watched in silence as her stern rose 20 feet into the air, paused momentarily and then, in one gulp, was gone, leaving only a small, dark opening in the ice. Within seconds, the pack had closed and nothing at all remained to mark where she had been.

They were 28 little dots marooned on the ice at the heart of the most hostile sea on the planet. They were utterly alone; the nearest outpost of civilization was many hundreds of miles away. Their last contact with the outside world had been more than 11 months earlier. They did not have a radio and nobody knew where they were. Their prospect of rescue was nil.

The *Endurance* sank at 5 p.m. In his tent that evening, Shackleton tried to describe in his diary what had happened. 'She went today,' he began. He struggled on for another 43 words and then gave up. 'I cannot write about it,' he concluded.

They expected to die, slowly and horribly. But what followed was the greatest Antarctic adventure there has ever been – and, arguably, the greatest story of human survival in recorded history.

The most remarkable thing about the *Endurance* saga is that they all lived to tell the story.

If Shackleton's objective was to cross Antarctica, ours is to locate his ship. We flew into Antarctica earlier today by jet from Cape Town. We touched down at a place called Wolf's Fang, about 500 miles from the coast in a region of the continent known as Queen Maud Land. From there we were flown by a ski-equipped turboprop aircraft to where we are now, a spot named Penguin Bukta on the Fimbul Glacier. It is perched by the very edge, where the ice ends and the sea begins. It's beautiful to behold but you can't help feeling a little vulnerable.

Soon we will set sail on the South African icebreaker S.A. *Agulhas II* for the Larsen C ice shelf on the other side of the Weddell Sea. The *Agulhas* is currently under charter to our expedition but normally she serves as a supply ship for the South African scientific bases on the continent, as well as certain regional sub-Antarctic islands. She will be packed. The crew numbers 44 while our team consists of 51 (29 scientists; six subsea technicians; seven data analysts and hydrographers; four documentary film crew; two administrators; one meteorologist; one medical officer; and an archaeologist).

The greater purpose of our expedition will be scientific, but towards the end we will plunge into the abyss in an attempt to find the *Endurance*. As a maritime archaeologist, I can say with little fear of contradiction that this will be the greatest wreck search there has ever been. Conditions and equipment allowing, we will battle our way through the ice until we reach what Shackleton himself called 'the worst portion of the worst sea on earth'. Then, once over Frank Worsley's legendary coordinates for the sinking, our unmanned search-and-survey vehicles will dive deep beneath the pack to 3,000 metres, where they will begin the hunt.

NOON POSITION: 70° 10.276' S, 002° 07.985' W

2 JANUARY 2019

New Year on the Endurance.
Awaiting the arrival of the scientists

How did we celebrate New Year on the *Agulhas II*? We enjoyed a barbecue on the helicopter deck and then went down to the ship's bar. Although this will be a dry ship, alcohol was served this evening as an exception. Contrast this with New Year on the *Endurance* exactly 114 years ago: on this day in 1915 the *Endurance* was, like us, off the eastern shoulder of the Weddell Sea. Despite being behind schedule, the expedition had covered 480 miles since entering the ice pack. It was, in fact, only 149 miles from the point on the coast where Shackleton planned to offload supplies and set up his winter quarters in a prefabricated hut – which, in some part, must still be in the wreck at the bottom of the sea.

On the morning of New Year's Eve, Shackleton's party had experienced their first real taste of the prodigious power of gathered ice when they were brought to a standstill by two closing slabs, each about 50 feet long and four feet thick, which caught the *Endurance* in a pincer movement and heeled her six degrees to starboard. To extract themselves they extended an ice anchor across the pack from the stern and then, by putting their engines to full astern and pulling in on the anchor, they were able to draw her to safety. The moment they were free, the two huge slabs that had held them slammed together and rafted up 12 feet over one another.

At midnight, the ship's bell was struck 16 times. The skipper of the *Endurance*, New Zealander Frank Worsley, joined Shackleton, Frank Wild and Hubert Hudson on the bridge, where they were soon united with the others. They shook hands, wished each other a happy new year and then went below to the wardroom, where they drank toasts to the King, the expedition and the success of their country at war. Wild called for three cheers for the Boss, which, noted Leonard Hussey (the team's meteorologist), 'caused Shackleton much embarrassment'. They then sang 'Auld Lang Syne' and retired to bed.

None of them could have had any inkling that by the next Hogmanay, the *Endurance* would be at the bottom of the Weddell Sea and they would be clinging to the ice for survival.

Our ship has her huge, overarching bow ploughed deep into the fast ice (a crust of ice that covers seawater but is attached to land – or, in our case, the towering shelf). We are awaiting the final wave of scientists who flew into Antarctica today by Gulfstream from Cape Town and are now at Wolf's Fang. Tomorrow they will continue to Penguin Bukta by a modified DC-3 prop-engine plane on skis.

This morning, the scientists already on board conducted field-work on the ice. Using a coring device and a probe, they measured electrical conductivity, salinity, temperature, depth and water pressure, as well as locking in samples of water to be analysed for biological and chemical activity.

On the bridge I met our two captains, Knowledge Bengu and Freddie Ligthelm. Captain Knowledge (as the crew call him) is Master and Captain Freddie is ice captain. Both were in their whites with gold epaulettes on their shoulders. They have been friends and colleagues for years and are clearly very much at ease in each other's company. Within the nautical pecking order they enjoy equal status and swap roles from voyage to voyage. Both are legendary ice skippers and, although they didn't say it, I have no doubt that they are as keen as I am to make history by finding the *Endurance*.

They introduced me to the mate, whom they addressed as 'Mr Mate', but I think his name is Jacques. He is clearly a character who says and does what he likes. I could feel him sizing me up. I don't know what to make of him but he clearly has the high regard of the two captains, so he must be good at his job to get away with so much lip.

Will we find the *Endurance*? I don't know; but in a few hours we will be on our way, and then a story will unfold which I know in my bones is as wide and deep and mysterious as the Weddell Sea itself. This is the kind of place where anything can happen, and you know it will.

NOON POSITION: 70° 10.290' S, 002° 07.898' W

3 JANUARY 2019

Football on the ice. Penguins stop play. The scientists arrive.
We proceed out of Penguin Bukta

Our departure has been delayed while we wait for the arrival of the remainder of the team from Cape Town. In happy tribute to the men of the *Endurance*, who enjoyed nothing more than what they called 'a bit of footer', our expedition leader, John Shears, decided that a kick-around on the ice would be a good idea.

Keeping the men fit, happy and motivated was important to Shackleton, so he encouraged football and hockey on the ice when conditions allowed. He himself was goalie. Their first game was on 20 December 1914, and was played between those who were supposed to be going ashore to make the crossing of Antarctica and those who would remain on the ship. The ship won 2–0.

It all began well enough for us. We had set up a small pitch on the low, flat expanse of frozen seawater at the bottom of the Barrier, the towering inferno of ice that marks the rim of the Fimbul Shelf, which runs some 125 miles up and down the coast. We were well away from both the ship and the wildlife. The day was spectacular. Acres of glistening ice, no wind, no snow and above us only sky: cloudless, blisteringly blue and radiantly sunny.

The football match, though, soon became totally shambolic. Naturally there was much slipping and sliding and a lot of angry-happy shouting. It was as if somebody had released the inmates of a madhouse and given them a ball.

When I first saw them, they were several hundred yards away and coming at us in a straight line: a troupe of jaunty little knee-high brushtail penguins. Surely, with so much space around us, they would deviate . . . but five minutes later they were suddenly storming onto the pitch. That, of course, was the end of our game.

We have been joined by the last members of the team and immediately we are under way for the Larsen C ice shelf on the east coast of the Antarctic Peninsula. Everybody has been out on deck. Antarctica may be melting, but you wouldn't know it here. The theatre about us is completely dominated by ice. Behind us is the Fimbul Shelf, its glacier trailing off eastwards as far as the eye can see. Ahead of us, to the west, lies an oceanscape padded with ice that eventually dissolves into the mist and becomes one with the sky. The water beside us is lit by bioluminescent plankton, a reminder that while there is little living above the surface, the ocean pulses with life.

But what most grips our attention are the huge flat-top tabulars, the aristocrats of the White Continent that crowd around our ship and make us feel small. When most people think of bergs, they envisage something resembling a jagged rock; and indeed that is how most of them are, or become. But not here, because this is where bergs are born. This is where they cleave from the tip of the ice shelf and either float off or sit for years with their bums buried in the mud. They are huge, vertiginous, brooding; not exactly hostile but certainly not to be messed with. They represent great forces of nature and time. If there is anywhere on earth more elemental than here, I cannot think of it.

NOON POSITION: 70° 15.363' S, 002° 42.103' W

4 JANUARY 2019

On Friday, 1 August 1914, the *Endurance* gave the first kick of her screw and slowly eased out of London's West India Docks into the Thames. One of the diarists described the large crowds that had gathered and how he had heard one bystander predict: 'Some of them will never see London again.' Three days later, on 4 August, Shackleton put into Margate on the Kent coast, where he read in the newspapers of the order for general mobilization. He returned to his ship straight away, mustered the team and informed them that he proposed to send a telegram to the Admiralty offering to place the *Endurance*, her stores and crew at the disposal of the nation. They all agreed, and Shackleton immediately cabled the Sea Lords in Whitehall.

The British navy was then the most powerful weapon the world had ever known and the man with his finger upon the trigger was none other than Mr Winston Churchill. Within an hour, they had received his response – just one word: 'Proceed.'

Fresh water has zero salinity and freezes at 0°C. Seawater, by contrast, has about 34 grams of dissolved salts per litre and does not begin to freeze until below −1.8°C. This is when the crystals begin to form. They rise to the surface, where they become slicks of a soupy consistency called frazil ice. With further freezing the crystals aggregate into pancakes with raised edges that can become compacted by winds and currents. At first these have diameters of less than four inches but, as they accumulate more frazil, they can grow to several yards across. The field we passed through was covered in patties the size and shape of large, carelessly thrown pizzas.

By afternoon we had left the broad band of mixed sea ice and enormous plateau-topped bergs that characterizes the coastal margins of Queen Maud Land at this time of year. Our speed was 12 knots. From the 'monkey island' above the bridge, where you can see for probably more than 10 miles in any direction, I counted over 70 icebergs, so this is not exactly open water. At 1000 hours everybody met in the ship's auditorium to be briefed on our objectives by the expedition principals. All are excited by the prospect of what lies ahead. Somebody asked how I rated our chances of finding the *Endurance*; I replied less than fifty-fifty.

I wonder what they would think if they realized that all this began over an exchange of ideas in the original Caffè Nero coffee shop on the Old Brompton Road in South Kensington. It was 16 August 2012, at 1145 in the morning, when I met with a friend to discuss the possibility of finding a couple of important historic wrecks.

The foremost of these was the *Terra Nova*. This was the ship that had carried Scott to Antarctica for his final, fatal assault on the Pole. It was the centenary of his death in 2012 and I had been asked by the Natural History Museum, which was planning a special exhibition on his life and achievements, if I could find the wreck, which had gone down off Southern Greenland in 1943. I had reliable coordinates and felt that, with the right equipment, I could find her within a matter of days.

While we waited for our coffees, I aimlessly leafed through a complimentary copy of *The Times* on the counter beside the till. The headline above a brief article caught my eye: '*Terra Nova* Found'. It came at me from out of nowhere, a perfect left hook. *Oof!* I was devastated. My friend asked what was wrong, and I told him. Almost without hesitation he said, 'Well, what about the *Endurance*?'

Ironically, I actually tried to talk him out of it, pointing out that the *Endurance* was under permanent pack ice in brutal conditions and that the technology was not quite ready for this kind of challenge. While there had been a little robotic work under the ice of the Arctic, that had only progressed several hundred yards – no more than lifting the corner of the carpet. What we were considering in that coffee shop was a bold new step forward in subsea technology and, I thought, probably a challenge too far. Thankfully my friend didn't listen and over the years a team was slowly assembled, along with the incredible technology that was necessary for the task.

NOON POSITION: 68° 07.III' S, 010° 12.839' W

5 JANUARY 2019

The Endurance: something more than a ship.
Out of the ice belt, but still surrounded by bergs

I have been directing underwater excavations and surveys since I was 28 years old. For 32 consecutive years I've worked on shipwrecks dating from antiquity to the modern day, often several in a single season.

In those early years my attention was generally focused on ancient wrecks, always in the Mediterranean and mainly off the coast of Italy. Field archaeology is all about the advancement and dissemination of knowledge, particularly new knowledge, so I would ask myself: what does this site tell us that we do not already know? If the answer was 'not a lot', then the site would be ignored; but if it raised important questions it would probably be surveyed and, depending on that evaluation, might be excavated. This yardstick has served me well throughout my career. The problem is that when you apply it to the *Endurance*, the site does not fare well.

If we discover the *Endurance*, she will not tell us much of any significance that we do not already know. Every detail of her construction has survived on paper and in the beautiful photographic record made by the expedition's intrepid Australian photographer, Frank Hurley, back in 1914–15. We know, too, what she contained and the circumstances of her loss; in this regard it is hard to think of a wreck that is better documented. So, from a strictly archaeological point of view, it is hard to justify what we are doing.

However, if we analyse it from a historical perspective, things look very different. Few, I think, would argue that this is not a site of outstanding cultural importance; the *Endurance* is to the Shackleton saga what the *Victory* is to the Nelson story. Both ships are rooted in the British psyche, both represent valour and all that is best in the human condition – but they go beyond that. They have become legends that belong to the world. One of the unpublished diarists on Shackleton's team said it well when he wrote that the *Endurance* was 'something more than a ship'.

In broad terms, we are trying to find the *Endurance* so that she might be protected into the future, when conservation science will have advanced sufficiently for a responsible body to consider raising her remains for preservation and public display.

Today was the anniversary of Shackleton's death. He died in 1922 of a suspected heart attack while on board his ship, the *Quest*, at South Georgia. At his wife's request, he was buried on the island in the whalers' cemetery at Grytviken. Earlier this evening some of us drank to his memory and then we all watched the excellent Kenneth Branagh film *Shackleton* in the ship's auditorium.

We've been bowling along at 18 knots. We are out of the ice belt but there are bergs everywhere, floating by like clouds. On the bridge all binoculars are out and scanning; not for big bergs but for what are called 'growlers': small, low slabs that are not easily seen but can inflict serious damage on a ship travelling at speed, as we are now.

Our estimated time of arrival at the Larsen C ice shelf is midday on 10 January. We released our first weather balloon.

NOON POSITION: 65° 57.721' S, 026° 15.203' W

6 JANUARY 2019

The ship's bell. Back in the ice fields

Ray is a rough, tough, spit-in-your-eye Texan with a physique straight out of Stonehenge. He pilots and builds underwater ROVs (remote operated vehicles). If we find the *Endurance*, Ray, like everybody else on board, wants me to raise something from the wreck, perhaps the ship's bell.

But it is not so simple.

Behind the planning for this project there was an Expedition Advisory Committee. The one topic we kept returning to was whether or not we should raise anything from the *Endurance*. There were a range of opinions but, in the end, we got bogged down in ethical issues. In addition, there were conservation concerns as well as legal questions regarding ownership (everything on the seabed is owned by somebody, it's just that usually they do not know it).

The whole matter became so complex that we finally decided nothing would be taken from the site. This is not to say that in the future, individual objects – and indeed the whole structure – should not be raised. There are items of information and cultural value on the site that, if left, will one day decay out of existence; but now is not the time. We simply are not ready for the reception and conservation of such artefacts. And so, for the present at least, the bell stays put.

But there is a question concerning the whereabouts of the bell, which does not appear in any of Hurley's deck photos. Some ships had a bell above the crow's nest so that the lookout could alert the steersman to any danger ahead, but on the *Endurance* we know they had a megaphone for this purpose. Some years ago, it was suggested during a symposium on Shackleton at Greenwich that the *Endurance* did not carry a bell. This is incorrect; there are fleeting references to it in the diaries and, indeed, it was even sketched by Walter How, one of the fo'c'sle hands (the fo'c'sle being an upper deck at the bow of the ship) and a decent artist. His drawing shows that the bell did not have the curves and proportions we normally associate with a traditional ship's bell of British manufacture.

There are, however, photos of their winter quarters that show part of an object with an unusual profile hanging above the table. It has been assumed that this was another light shade, but comparisons with How's drawing confirm to my satisfaction that it was the bell. Clearly it had been brought below deck for winter. The purpose of the bell was mainly to alert other ships to their presence during fog and to mark the passage of the watches, which ran from one to eight bells. Once the ship was icebound in the Weddell Sea and without regular watches in place, the bell was brought below deck – where perhaps it was used to summon people to eat, as Shackleton always demanded punctuality at mealtimes.

I had a meeting with Captain Knowledge in his cabin. Together we looked at the Copernicus Sentinel-1 satellite images of ice conditions around the huge slab of ice that calved from the Larsen C ice shelf. That is our next destination. We are experiencing the best ice conditions for many seasons and he thinks we have a good chance of getting in close to what is left of the shelf.

For much of the day we have been in close pack ice. The ship is on manual and we are finding open water wherever we can that will take us in the right general direction. There are 750 miles to go before we reach Larsen C. Our current ETA is midday on 10 January.

NOON POSITION: 64° 11.611' S, 036° 41.319' W

8 JANUARY 2019

Macklin's lost diary. 150 nautical miles from the tip of the Antarctic Peninsula. Wildlife becoming richer

Nothing beats a good diary.

The real red meat of what happened down here in the Weddell Sea a little over 100 years ago is to be found within the diaries of those who were on board. Many people only read Shackleton's book *South* (not to be confused with his diary) – which is an excellent book, but it is Shackleton burnishing his legacy. If you want to know what was really being thought, said and done you must go to the 'I was there' accounts; but – and here we come to the problem – only three of the nine diaries have been made fully available through publication.

As might be expected, some of these are more revealing than others. Frank Worsley has a lot to say – but he was a romantic who was writing to be read, and liked to embroider for the sake of the story. Thomas Orde-Lees's journal is nicely descriptive; James Wordic is a bit too guarded and discreet; Reginald 'Jimmy' James is good on science but otherwise a bit bloodless; Harry 'Chippy' McNish is unpunctuated and stripped-down terse (but I love him); and Shackleton is probably the most irritating of all because he, potentially, has the most to say, but he does not say it. And more than that, he simply will not allow you into his head.

My favourite is the diary of the physician, Alexander Macklin, which survives in the archives of the Scott Polar Research Institute (SPRI) at Cambridge. Of all the diarists he is easily the most thoughtful and erudite. Unfortunately, only the second half of his diary survives – the first half went down with the *Endurance*. Macklin's cabin was situated on the starboard side of the vessel beneath the poop deck and thus became one of the first to be flooded. He made several attempts to rescue his belongings, but soon the water turned to frazil and then froze. And so his diary was lost with the ship – something which remained with him as a source of grief for the rest of his life.

We are now 150 nautical miles from the Antarctic Peninsula and closing. We have fair skies and unfurrowed seas.

Ice that survives to the following year is called 'first-year ice', while that which survives two summers is called 'multiyear ice'. Up until now we have mostly been passing through young ice, but today we've been shouldering our way through mixed first-year and multiyear ice, some of which is over three metres thick and requires utmost vigilance even though we are still in what is called the Marginal Ice Zone. Apart from the floes, there are large bergs of monumental proportions dotted around in every direction as far as the eye can see. On the bridge there are several people constantly scanning with binoculars, looking for ways through but also making sure we do not whack any growlers.

Since yesterday the wildlife has become noticeably richer. Occasionally I spy a forlorn emperor penguin on a floe, but the Adélie penguins are now ubiquitous. I have also seen the odd crabeater seal hauled up on the ice, and today I saw my first Weddell seal. As for what's happening overhead, boy, do we have petrels: snow petrels, Cape petrels, Antarctic petrels and giant petrels. During dinner, when somebody learned I was interested in the birds around these parts, he described for me one he had seen as all white with a black cap – that can only be an Antarctic tern.

Later tonight we will pause to make two CTD (conductivity, temperature and depth) casts and put out the bongo nets. A CTD cast involves sending a probe to the bottom of the sea on a spooled line, generating a range of evaluations – salinity, water density, sound velocity, etc. – all of which are essential for the calibration of the scientists' equipment. The bongo nets are simply a pair of long, funnel-shaped, fine mesh nets resembling bongo drums which are used for the collection of marine biota, in particular phytoplankton.

NOON POSITION: 62° 37.194' S, 048° 41.361' W

10 JANUARY 2019

Where are the whales? Back among the bergs. Marine acidification. Passing the Larsen B ice shelf

I saw a minke whale today. Two days ago we saw another. But, as far as I am aware, that's it. Where, I ask myself, have all the whales gone?

I decided to look back to see what Shackleton was observing in the way of whales at this time. On this very day in 1915, the day in which he made landfall with the continent, he wrote: 'Loose pack stretched to the east and south, with open water to the west and a good watersky . . . the *Endurance* continued to advance south. We saw the spouts of numerous whales and noticed some hundreds of crab-eaters [seals] lying on the floes . . . a few killer-whales with their characteristic high dorsal fin also came into view.'

Glancing back to his entry for the previous day, I found the following: 'Two very large whales, probably blue whales, came up close to the ship, and we saw spouts in all directions.'

These are the great feeding grounds to which the leviathans used to migrate every summer. Right here is where the greatest beings on earth – the 100-foot blue whales – would come to scoff. The largest population of this species in the world was based in Antarctica. But where are they and all their cousins now?

Shackleton's marine biologist, Robbie Clark, a dour and hard-working Scot, was particularly interested in the whales and troubled by the rampant over-exploitation of their stocks, in particular that of the humpback whale. After the *Endurance* left South Georgia for the Weddell Sea, Clark noted only two sightings of humpbacks. He later concluded that because of their unrestricted slaughter, 'the humpback stock is threatened with extinction', and urged international protective legislation. Yet despite the statistics and obvious diminishing returns, the whaling industry was allowed to stagger on until the mid-1960s. In recent decades whales have been making a comeback, and 15 years ago they were again a fairly common sight around here. I know because I was in these parts over a decade ago and witnessed them for myself.

So again I wonder: why it is that we are not spotting them now? To my simple mind, if the whales are not here then their food is not here. Today we are merrily chiselling away at the very foundation of these fragile food webs, and this has ramifications that go beyond horrific.

Off King George Island, just a short distance around the corner from us, the krill biomass that the whales feed on has declined by a staggering 80 per cent and is still falling. This can be linked directly to the loss of the ice pack down the west of the Peninsula, because juvenile krill live within the loose crystalline latticework on the underside of the ice.

And what about the penguins? In the same general area on the western Peninsula, in the last 35 years, the breeding population of Adélies has also declined by roughly 80 per cent. This too can be attributed, both directly and indirectly, to climate change – the dramatic reduction in sea ice, the decline of the krill, as well as the increased snowfall (a result of the warming air over the Peninsula) that leads to melt-water which, in turn, chills the eggs and drowns the chicks. Whole colonies have literally been wiped out, leaving what are called 'ghost rookeries'.

And finally, what about those 'hundreds of crab-eaters' that Shackleton saw on this day in 1915? Crabeater seals don't actually eat crab; they eat krill. How many crabeaters have we seen in the past 10 days? I have counted precisely six.

They say the Antarctic Peninsula is the canary in the coal mine for climate change, global warming and ocean acidification, because everything down here is in such a delicate state of balance. Temperatures have already risen by almost two degrees. The acidification of the sea has been relentless in its climb, because the oceans exchange gases with the atmosphere in such a way that all the pollutants we put up there eventually end up in the sea.

I am one of a very few people who have had a keyhole-sized glimpse into the future of marine acidification. I have seen the horrors it can bring. Between 1986 and 1988 I directed the excavation of a wreck off Panarea, one of the islands in the Aeolian Archipelago to the north of Sicily. Known as the Dattilo wreck, it contained a rare consignment of beautiful, black-painted fineware (now on display in the Archaeological Museum of Lipari) which was of outstanding archaeological importance because it came from the Classical period of ancient history – a sunburst moment in the annals of human achievement that gave us Pericles, Plato, Aristotle and the Parthenon. Very few wrecks are known from that era and none with such an elegant cargo. I shall never forget my first sight of those cups, dishes, plates, jugs, beakers, bowls and oil lamps all jumbled about upon the seabed. It was as if the gods of the deep had been sitting down to dine when something happened that made them spring to their feet, but in their haste they knocked over the table and sent the crockery sprawling.

Normally a shipwreck is a haven for sea life, an oasis for anything that can burrow, creep, crawl or swim. Over time it evolves into a balanced association of eating guilds unique to that wreck, in which everything is fodder for something else. But not on the Dattilo wreck. Although it was full of nooks and niches, all perfect for the accommodation of little aquatic creepy-crawlies, there was nothing. It was a dead zone. No vegetation, no corals, no sponges or other squidgy things, not even barnacles, crabs or flitting fishes. It was like a lunar landscape, a desert of brown granular matter punctuated by occasional rock outcrops.

There was something else very freaky about the site. Wherever one looked, there were bubbles issuing from the seabed; most commonly they trickled up in single ribbon-like streams, but elsewhere they tumbled surfacewards in great gushing bursts. In places where the vents were more linear, they rose in curtains. As all divers know, bubbles normally get bigger as they rise, but not here – here, they diminished. We guessed that they were of a gas that was dissolving, but beyond that we were mystified.

And there was also another phenomenon I had not met before: after we had worked on the site for 30 minutes or so, our lips began to tingle. This continued for several hours after we returned to the surface. We blamed it on what we called no-see-ums – tiny jellyfish, or medusae, barely visible to the naked eye under water.

In the end, all this was explained by two visiting volcanologists from the University of Palermo who told us that we were working within the crater of a live, submerged volcano – part of the Vesuvius, Stromboli, Mount Etna nexus. The gas was mainly carbon dioxide, which was dissolving into carbonic acid; it was this which had killed, or driven off, all the sea life in the vicinity and made our lips sting. At the time I had never heard of ocean acidification, but now I wonder if the Dattilo effect was not a microcosm of the changes that are to come.

There is little we can do to moderate the effects of increasing acidification. The amount of time needed to undo the damage would be considerably greater than the time it has taken to turn the ocean – this seminal, life-rejoicing medium – into a heinous pisspot of life-gagging pollutants. We discharge more than seven billion tons of carbon-containing gases each year into the atmosphere. Even if we were able to halt it all tomorrow, scientists tell us that it would still take tens of thousands of years for the pH of our seas to return to pre-industrial levels.

This continent has taken a terrifying lurch into the unknown. Antarctica is in pain, and it is we who have unleashed the fiends. Whether it is carbon pollution, climate change, ice loss, ozone depletion, ocean warming, rising acidification or the prospect of mass extinctions, it is humans who have done the hellish thing and shot the beautiful bird.

That is why the science that we are doing down here is so vital.

We are now picking our way carefully – very carefully – down the eastern side of the Antarctic Peninsula.

At 0600 hours the whole ship was jarred when the *Agulhas* had a go at a large ice floe that was more resilient than it looked. The ship backed off and found a way around it, but by then everybody was awake. Because of the jolt, some of the scientific kit registered malfunctions. 'Electronics don't like that stuff,' summarized Ray at breakfast.

We should reach the Larsen C ice shelf during the early hours of tomorrow morning.

NOON POSITION: 65° 20.136' S, 059° 27.715' W

11 JANUARY 2019

We touch the Larsen C ice shelf. Shackleton's 'fateful day'.
Arriving in the OZ

I have never seen the swallows return to Capistrano, nor have I
been to Connecticut in the fall. I have never witnessed the Northern
Lights, nor have I seen Paris in the spring. But last night I reached
out over the very tip of the bow of the ship, extended my forefinger
and touched the Larsen C ice shelf – and there cannot be many in
the world who can claim that.

It was 0230 in the morning; the sun was red-rimmed and low
in the sky, but still shining. The ship was asleep. At the very bow
it was just me, the South African scientist Annie Bekker, and a
member of the crew. On the bridge the captain was trying to find
open water beside the shelf so that we could test the ROV under
the ice. To maximize the clear water off the stern, we nosed right
up against the shelf so that our bow was actually touching the
barrier. Soon, however, loose floes brought in on the wind began
gathering about us, and we backed off in search of somewhere
more sheltered. As we drew away I noticed a small red smudge on
the ice wall: we had accidentally left a little of our paintwork behind,
just enough to say we were here.

The ice conditions have been kind to us. Just last season a British
Antarctic Survey icebreaker, the RRS *James Clark Ross* (named after
the British explorer who was in the Ross and Weddell seas during
the early 1840s), attempted to break through to the Larsen C and
giant iceberg A-68 but was thwarted by four to five metres of ice.
They were obliged to turn back. Even if we do not find the
Endurance, we have already done well.

And so we have arrived. It has taken 10 days, 15 hours and 20
minutes, but we are exactly where we want to be, between the

Larsen C ice shelf and Iceberg A-68, the massive 200-metre-thick, one-trillion-ton berg that calved from the shelf. Equipment trials begin later today. We are so keen to get started, it almost hurts.

I love a grand gesture. For their intended journey across the ice after abandoning the *Endurance*, the crew were obliged to ditch all unnecessary weight. Shackleton cast his gold onto the snow and challenged everybody else to do likewise.

Orde-Lees described the exercise in his diary: 'Our leader proceeded to set an example by deliberately throwing away all he possessed – away went his watch, about 50 golden sovereigns . . . books and a dozen other things, whereupon we all did likewise until there was a heap of private property probably of some hundreds of pounds' value, lying about all over the floe.'

Shackleton called it 'The fateful day' – 27 October 1915 – the day they abandoned ship. That evening he wrote: 'The end of the *Endurance* has come . . . [she] is crushed beyond all hope of ever being righted, we are all alive and well . . . the task is to reach land.' But what land? Where were they heading?

During the last 24 hours I have not been able to stop thinking about Shackleton's intentions because yesterday we passed Robertson and Snow Hill Islands, and the day before, Paulet Island. Many months ago I had a gentlemanly difference of opinion with my colleague Toby Benham of the SPRI over where Shackleton was heading after they had abandoned ship. Toby said Robertson Island; I said Paulet or Snow Hill. I argued that Shackleton had never mentioned Robertson in his book *South*, only Paulet, with a sideways reference to Snow Hill. However, since then I have done much more reading of the diaries and can now see that we were both correct. Despite what Shackleton later wrote, there can be no doubt that, on the 'fateful day', he had been thinking of Robertson as much as Paulet.

On 27 October, the day they left the *Endurance*, one diarist wrote: 'By 8.20 p.m. everything necessary for our proposed sledging journey to either Robertson Island, Sugar Hill [Snow Hill] or Paulet was out on the floe . . .' Three days later, on 30 October – five minutes after shooting Chippy McNish's cat, Mrs Chippy – they set off for the Antarctic Peninsula.

In *South* Shackleton only mentions Paulet, which, at 346 miles away, was the nearest point where there was any possibility of food and shelter. A stone hut had been built there by Otto Nordenskjöld's 1902 Swedish expedition after they lost their ship, *Antarctic*, off the island. Shackleton knew that the hut was full of supplies because he had bought them himself in London when he had been asked to equip a relief expedition.

So why didn't Shackleton mention Robertson Island in *South*? The answer must be because going there would have been a serious error of judgement, since it held neither food nor shelter. As the foremost British explorer of his day, a role he much cultivated, he did not want that on his record. Even Shackleton was fallible.

We have arrived in the OZ, the Operations Zone. There are feelings of elation throughout the ship. Now the serious work begins.

In a corridor earlier I passed Claire Samuel, who is in charge of all back-deck ops. She and I have collaborated on a range of deep-ocean projects around the world, so we know each other well. We tend to hold our ad hoc sessions in the doorways of each other's cabins and we always say exactly what's on our minds – today, though, we just exchanged nervous smiles. We both know that from now on it's going to be one problem after another, one crisis after the next. And in the end – win or lose – we are both going to feel ourselves completely wrung out. We have been here before.

Surrounded by many floes and large tabular bergs, before anything else, we need to stress-test our subsea vehicles. Finding the clear water necessary to conduct trials has not been as easy as we thought; nonetheless, by 0930 hours we had the ROV in the water and ready for its first dive. There was a tense moment when the umbilical, which energizes the ROV and connects it to the ship, came into contact with a floe and was dragged under the stern. Fortunately there was no damage. While we are usually stationary, the floes are being carried along on a 1.4 to 2.0-knot current and sometimes we have to push them away using the ship's FRC (fast rescue craft). Once the ROV had gone through its buoyancy and systems checks, it then placed transponders on the seabed at 500 metres for HiPAP (high-precision acoustic positioning) calibration. The HiPAP allows us to communicate acoustically with the underwater vehicles by way of a transducer at the end of a shaft which we lower into the water through the moon pool, an opening through the bottom of the hull that gives access to the water below. With the first trial successfully completed and our acoustics in order, we need now to find a patch of open water beside the Larsen C shelf where we can conduct the ROV's first under-ice test dive.

Although we will be conducting science until 24 January, I am now monitoring conditions over the *Endurance* wreck area with our ice skipper. Unfortunately, there is still dense coverage over the site. The weather so far has been too kind to us. Basically, we need something to happen – something that'll grab that pack by the scruff and shake the hell out of it. In the long history of seafaring, I must be the only person who has prayed for a thumping good storm.

NOON POSITION: 66° 20.921' S, 060° 18.611' W

12 JANUARY 2019

We touch giant iceberg A-68. Thwarted by the floes. The mate

Doubts have been expressed by some on the ship as to whether volume one of Macklin's diary, which went down with the *Endurance*, would have survived.

I grant that it will have suffered some deterioration and will be distinctly mushy but I am absolutely confident that if the wreck is in semi-intact condition, the diary will still be there – and in large part legible.

Water, you see, is a surprisingly good preservative, and it's even better if temperatures, light and oxygen are all low. Take the *Titanic*, for example, which went down in 1912, just three years before the *Endurance*. Many written items including books, banknotes, calling cards and even pieces of newspaper have been salvaged from the wreck. People are often amazed that even after 75 years on the seabed, they remain in a legible state.

At 2001 hours our bow touched A-68 for the first time. Others have tried, but we are the first to reach it – a historic moment.

It has been a difficult 24 hours that has left everybody exhausted. We are two days into our mission and already we are all on our nerves. When we started the day right up against the Larsen C ice shelf we were hoping to conduct equipment trials, but because of unfavourable currents and a persistent north-easterly wind that drove the floes towards the shelf, the sea was too congested for subsea ops.

So we moved some 12 miles south along the shelf towards an area that, from the ship's ice radar, seemed to be more sheltered – but again we were thwarted by the floes. All evening we struggled with them; we tried driving them away with prop wash (the turbulent water created by the propeller) and shoving them to one side using the FRC boat, but nothing worked. After midnight, we drew off from A-68 and headed out to find an uncongested spot away from the berg in which to conduct further CTD and coring work. Throughout the day much use was made of drones to search for openings and to monitor the drift of the large floes.

I meet regularly with the ice captain, Freddie Ligthelm, in a quiet corner at the back of the bridge. Our main focus, of course, is the pack over the *Endurance* search area, but today we also studied our current location. A-68 is still dragging one end of its keel in the mud. Originally this was its northern extremity; however, in recent months the berg has swung around 180 degrees so that what was its northernmost tip is now its most southern. What this signifies for us is that A-68 is at present aligned in a 'five minutes to eleven' position, creating a confined V-shaped area of sea space between the berg and the northern edge of the Larsen C ice shelf. We are in the middle of this area. The two converging sides of the V, however, do not quite connect, leaving a bottleneck of just a few miles' width. The Weddell Sea gyre moves in a clockwise direction, and that passage has become a choke point for floes and bergs trying to nudge their way up the coast on the current. It is as if all that ice is trying to pass through a funnel in reverse, but once through, it all just spews out into the zone where we are now.

I put it to Freddie today that we are not going to get through that gap. 'No,' he replied slowly, 'unless something changes, it's just too dangerous.'

'It would be a bit like trying to navigate the Hoth Asteroid Belt with everything just coming right at you,' I said. Rather pleased with my little analogy, I watched his face for a response and when I didn't get one I was for a moment tempted to explain, but then I thought, 'Nah, perhaps he doesn't do *Star Wars*,' and let it go.

One of our main scientific objectives is to reach the 2,246-square-mile area that, until recently, has been cloaked by A-68. It is an important sector for scientific evaluation. The seabed communities within that tract are undergoing regime change; the old residents are suffering, their habitat and food supply chains are in disarray, and new invasive species are moving in to trample and gorge. It is important, therefore, to study the various processes of change taking place before everything has been assimilated. It is only 50 miles to the south of us – so close! But as Freddie and I looked at the satellite imagery of where A-68 had been before it cleaved, it seemed to me that ice conditions in that patch were far worse than where we are now, and even if we could get through we might not be able to conduct a marine biological survey.

I mentioned this to Freddie and he observed that A-68 seemed to be acting a bit like a barrier: everything was backing up behind it. 'As it is now, we're not going to make it. It would be like trying to wade through cement, and twice as dangerous,' he said; and then, without missing a beat, 'I'm not Han Solo and this isn't the *Millennium Falcon*, you know.'

In my cabin afterwards it occurred to me that, just as we are not going to make our final destination, nor did Shackleton make his. But – and here's the thing – both of us got to within one day's sail of it.

I finally spoke to the mate today. He is disciplined, hard-boiled (the full 20 minutes) and mostly unsmiling. On my first day I was told that he didn't suffer fools or academics, so I have been keeping my distance. As a seaman it is obvious that he is respected by all and the cadets love him as a teacher. As I watch him conning the ship, it strikes me that he is not of our era. He should be behind the wheel of an old square-rigger battling to windward off Cape Horn with ripped canvas aloft, buried decks below and white water gushing from the scuppers. Anyway, today he spotted me on deck with a vintage sextant of the same type and model they had with them on the *Endurance*. Interested, he came over and for a long time we talked about navigation. I think he may be the only one on board who was brought up on traditional methods of position-fixing and chart work. I wouldn't say we are now bosom buddies – he's not that kind of a guy – but at least he's no longer looking at me as if I'm just another dozy academic with pig shit for brains (which, I believe, is how he referred to me soon after I joined the ship).

NOON POSITION: 66° 46.511' S, 060° 00.074' W

13 JANUARY 2019

A wooden ship will always deteriorate on the seabed. There will be chemical decay, bacterial attack and a range of mechanical forces at work. However, the purity of the Weddell Sea, its deep cold, moderate bottom currents and the absence of light at 3,000 metres, should inhibit the ability of these agents and factors to inflict ruin on her timbers.

Potentially more worrying is the dreaded shipworm, or *teredo navalis*: a wood-consuming marine parasite that is to wooden ships what moths are to cardigans or deathwatch beetles to timber-frame houses. These creatures enter the wood as larvae about the size of pinheads, but then metamorphose into pallid reddish 'worms' that turn into the grain, and then swell quickly to the width of one's small finger while growing in length to 20 inches or more. My early work was on ancient ships in the Mediterranean where the hulls and upper works had all been gobbled away; the only timbers to survive were those that went into the seabed. The sub-bottom is essentially anaerobic (or oxygen-free) and here the *teredo*, and gribble worms in general, cannot survive.

The happy news is that studies conducted in the 1980s and again more recently have confirmed that wood-consuming marine parasites do not exist in Antarctic waters. But having said this, in recent decades the *teredo* has become established in Port Stanley harbour in the Falklands, where I have found them on 19th-century wrecks.

It is also possible that the hull of the *Endurance* became infested with marine borers when she passed through the tropics so that, when she sank, she took with her to the seabed the seeds of her own future disintegration; but this is highly unlikely as her hull was sheathed in greenheart, an exceptionally hard wood that was considered to be largely *teredo*-resistant.

Another day of frustration. The ship is again trying to create space for underwater operations by driving away the floes using her stern thrusters or fast rescue crafts. I overheard the mate say to someone on the bridge that since we got here, we have just been 'pinballing around in search of open water'. Nonetheless, overnight we did manage to conduct work at three CTD/coring stations.

At midday, the large Eclipse ROV was lowered over the stern and released for its first under-ice trial. Deck to deck it was gone two and a half hours and traversed about 190 metres horizontally under the ice. There were some buoyancy issues, and a couple of jubilee clips snapped because of the cold, but otherwise it passed without incident. However, 190 metres is as far as it can go, and this, I cannot deny, is disappointing. It seems the ROV team were not briefed by the scientists that extreme horizontal deployments might be part of their mission. There was tension in the air when this became clear, and things were said. My only interest in the ROV is for vertical wreck-study dives so, thankfully, this one has nothing to do with me.

For the first time we also deployed the brand new HUGIN Autonomous Underwater Vehicle (AUV) No. 9, specially designed with an upward-looking sonar for under-ice scrutiny. Some worrying buoyancy problems were experienced; simply put, we could not get its nose down low enough to dive. In flat calm conditions several dives were attempted without success. Ballast adjustments were made and the propeller RPM was increased, but without result. We are now consulting with the manufacturers, Kongsberg of Norway, who have one of their senior technical staff with us on the expedition. This, of course, is what trials are all about and we have no doubt that these issues will be resolved. We will conduct more trials tomorrow.

NOON POSITION: 66° 29.838' S, 059° 59.189' W

14 JANUARY 2019

The sinking of the Endurance. *Biological dive by the ROV.*
More sea trials of our new AUV

The three conditions that probably most influence a sailing vessel
as it sinks through the water are ballast, shape and drag. On sinking,
most ships tend to descend keel first or somewhat down at the
bow. In four decades of looking at shipwrecks, I have only seen
two wrecks upside down on the seabed – and in both cases they
had left the surface in somewhat unusual circumstances and were
in relatively shallow water, so that they did not have time to establish a level of equilibrium as they were going down.

With a sailing ship the masts, yards and canvas will always
impose a significant element of drag upon a sinking hull, normally
ensuring a mainly upright or down-at-the-bow plunge. We have
seen in Hurley's photographs of the *Endurance* that her masts were
not upright when she made her final slide; we can also see that her
poles are still roped to the hull by both standing and running rigging.
As she sank through the water there would have been an incredible
trail of masts, yards and canvas behind her, creating a much greater
drag than in a normal sinking scenario.

As long as the *Endurance* left the surface in a reasonably
coherent state, though, I believe she would have landed keel first
or down at the bow – and that, although she may have heeled
somewhat, she most likely came to rest in an upright or semi-
upright position. However, there is another factor involved which
takes her outside the normal range of my experience, and that
is her extreme depth.

The sea is perfectly calm, the temperature –4°C. This morning we began a ROV dive to 400 metres' depth off Cape Framnes, for the biologists to study marine communities on the seabed that until some decades ago were tucked away beneath an ice shelf. Video of this was relayed in real time to the lounge, where it was watched by most who were at that time off duty.

Once the ROV was back at 1830 hours, CTD and coring work was conducted at Station No. 2, following which AUV 9 was launched for further tests. Once more it had trouble while submerging and had to be recovered for further ballasting. There is now little doubt that the next trial will be successful, so a mission profile has been prepared for its first dive under ice.

At 2230 hours the *Agulhas II* went alongside an enormous floe, over four miles in length, to prepare for the under-ice AUV trial.

We are now concluding the trials phase of our programme, and until 24 January we shall be concentrating on scientific data-gathering.

NOON POSITION: 66° 01.302' S, 060° 20.637' W

15 JANUARY 2019

Endurance upon the seabed

The overwhelming majority of shipwrecks studied by archaeologists are within a depth of 50 metres and are thus reachable by aqualung divers. Shackleton's ship, by contrast, is 60 times deeper, at 3,000 metres. The *Endurance* takes us into the archaeology of hyper-depth, of which there is exceptionally little experience, and thus we have hardly any data upon which to draw for guidance.

There is reason to believe that the dynamics of sinking ships in these depths are different from those experienced by other wrecks. In planning for this expedition, I have had to consider what actually happened when the *Endurance* hit the ocean floor. Was contact gentle or not? Did she retain her form or did she break open? Fluid scientists love their tidy worlds of perfect spheres of known mass in liquids of known viscosity, but the plummet of Shackleton's much-injured ship would have been anything but perfect.

The ship was heavily timbered and exceptionally well fastened; at the time of her construction she was called the second strongest wooden ship ever built (the first being Roald Amundsen's *Fram*). Although her hull had been ruptured in places by the ice, and indeed by Shackleton's men, before she sank, I still believe she would have retained enough structural integrity to survive impact in a reasonably coherent state. Of the wooden wrecks I have seen, most of their hulls had opened up, but this had nearly always happened well after the vessel came to rest on the seabed.

The only wooden ship in truly deep water that I have been able to study in detail was completely different from anything I have ever seen within standard diving depth. I was astounded by her.

She was a three-masted barque, about 170 years old, in approximately 5,500 metres at the bottom of the Southern Ocean. Her bottom keel had hit the sea floor first, with such violence that it had blown open the hull into three longitudinal sections. Her entire cargo had spilled out onto the seabed. I had never seen anything like it before – a dense, oval-shaped field of mixed 19th-century artefacts.

If we succeed in locating the *Endurance*, we may see a similar picture, but the greater likelihood is that we will find her hull to be roughly upright and still semi-coherent. I base this view on a lifetime spent in wreck archaeology, as well as my knowledge of her specially braced construction and the virtual certainty that she went down keel first.

In the spirit of Scott (but not of Shackleton), the specialists on board have been taking turns to give lectures in the ship's auditorium on their respective areas of expertise. One of the most important talks was delivered by Katherine Hutchinson, a young South African scientist whose area of focus is an interoceanic circulation system that is of fundamental importance to the health of the planet. That system begins right here where we are now, in the relatively pack-free areas beside the ice shelves of the Weddell Sea.

In her talk Katherine explained how, since around the middle of the last century, circulation movements within the Atlantic Ocean have slowed down by 15 per cent – and, for reasons not properly understood but which at some level must be linked to the global conveyor belt, the Indian Ocean has experienced a massive heat gain of one degree in 20 years. If these trends continue, Katherine said, they will have profound consequences for the world's climate and, by extension, the ability of its myriad life forms to feed themselves. We are, she reminded us, already seeing the consequences of climate change in extreme storms, heatwaves, droughts, wildfires, floods, crop failures, defunct fisheries, monsoon volatilities, disrupted rainfall patterns and so on.

One of the main contributing factors is the destruction of the ice caps because of warming caused by the increase in greenhouse gas emissions. The reduction in ice over Greenland has been accelerating for decades, to the extent that it is now losing an average of 260 billion tons a year. The situation in the Antarctic is no better.

The Antarctic ice sheet, which in places is three miles thick, covers 98 per cent of the land mass and contains 61 per cent of all the fresh water on earth. Because of warming, however, many of the continent's most important glaciers are now in a state of rapid decay. Eight major Antarctic glaciers are now retreating at a rate of over 400 feet a year. Some, those that spill into the Amundsen Sea on the continent's western coast, are retreating at a staggering 600 feet a year. A recent satellite investigation by glaciologists at Leeds University found that global ice loss has risen by 65 per cent over 23 years – two-thirds of that loss has been caused by atmospheric warming, with the rest attributed to rising seawater temperatures. In short, ice loss is now following the worse-case scenario envisaged some years ago by a United Nations panel of experts. The fear is that if fresh water continues to intrude upon interoceanic circulation at its current rate, the entire system will abruptly collapse with unpredictable, but certainly dire, consequences for the planet.

I have been reviewing the seabed footage from yesterday's ROV biology dive. Most of the life we observe is pretty sedentary, or what we call sessile, meaning it does not move (sponges or barnacles, for instance). Our problem is that anything that swims tends to flee as soon as it sees us. But in the video, much to my delight, a quietly subversive little fish suddenly appears from out of the gloom, pauses, and then swims over and nibbles the corner of our lens.

Frank Hurley, in his diary for this day, also recorded a rare encounter with marine life: 'We witnessed a phenomenal sight. Hundreds of crabeater seals speedily made their way towards the ship, and treated us to a wonderful display, gambolling, sporting, racing and diving under the ship.' He caught all of this on film and included it in the movie he later made, which created a sensation wherever it was shown.

NOON POSITION: 66° 00.817' S, 060° 21.627' W

16 JANUARY 2019

The loss of AUV 9

Last night, not long after midnight, we lost AUV 9. As I write we still haven't found it. AUV stands for Autonomous Underwater Vehicle – but AUV 9 is more than that. It's brand new, unique, worth squillions, and was especially designed for this expedition.

When you are kicking at boundaries, things can go wrong. To paraphrase Shackleton's favourite poet, Browning, our reach sometimes exceeds our grasp. Also, of course, we are deep into the most hostile body of water on the planet – we are down in the very basement and, what is more, we have kicked away the ladder. When a crisis occurs in the Weddell Sea, you cannot just reach for the phone and call for help. Here we have to look to ourselves.

It was 0200 hours, the middle of the night, but this being the austral summer it was daylight outside with the sun just a few degrees above the horizon, casting long, inky shadows across the floes. Down in the ops room there was quite a bit of nervousness, as we were on the threshold of a historic moment: AUV 9 was about to embark on its maiden voyage. Its objective was to study the submerged surface of an ice field and it had been created just for this job. At huge cost, it was manufactured by Kongsberg for Ocean Infinity, the foremost deep-ocean search-and-survey company in the world.

The buoyancy problems we had previously experienced with AUV 9 had all been resolved, and now it had gone through all its pre-dive checks without any irregularities. Nonetheless, we were operating on the outer fringes of what was technologically possible and we were not completely confident it would perform as instructed. Because of that uncertainty, this was going to be a relatively short trial of just three hours beneath a very large floe. What concerned us was that once the AUV had entered the under-ice zone, it would be in unsupervised mode and out of communication. If something went wrong, we would not know it.

There were six of us in the ops room at the time: Claire Samuel, myself, and four guys from the robotics company Ocean Infinity's top-gun AUV wing. On the bridge was the AUV team supervisor, Channing Thomas. So far, the launch seemed to have been successful: the AUV had performed a fairly steep descent to its programmed 'loiter' station (a notional box 100 by 100 square metres), where final systems checks were conducted, after which it was released from supervised mode to begin the mission.

At precisely 0225 hours, it slipped under the ice – and almost immediately we lost comms. For 37 long minutes, we waited. Conversation was stilted; everybody was tense. I thought of Mission Control in Houston when Apollo 11 passed round the dark side of the moon and there was nothing they could do but wait for it to reappear.

Precisely on schedule, at 0302 hours, AUV 9's return vector line on the screens suddenly turned yellow, informing us that contact had been resumed. There had been a tiny bit of drift, which was to be expected, but the vehicle had followed its operational profile and was heading towards the loiter area. Once there it would go into a holding pattern while it was debriefed and then re-mandated before embarking on the next phase of the mission, which would again take it back under the ice. Aware that this was a historic moment, one of the AUV team, Espen Strange, clicked a photo of us all in a smiling huddle around the computer screens.

An instant after that photo was taken, everything seemed to crash at once. Lights flashed, screens registered behavioural problems, somebody said 'Uh-oh' – and then, from 260 metres down, AUV 9 began a climb to 167 metres. Clearly it had taken matters into its own hands and decided to abort the mission. In itself, this was not so unusual; hugely irritating, certainly, especially when things were going so well, but nothing the AUV team hadn't dealt with many times before. Although it was now self-piloting we did not think that it would deviate far from its previous heading and we therefore expected it to surface away from the floe, in water that was largely open. On the bridge, the boat crew was scrambled to perform what was expected to be a routine recovery. At the door I passed my old friend Chad Bonin, a cheeky, charismatic Cajun from the Louisiana Bayou. He was heading off to get kitted up in his polar gear to oversee boat operations.

'Wanna come?' he asked.

It's a young guy's game, I thought to myself. 'No,' I said, 'I'm off to bed.'

The next thing I knew was when somebody thumped on my door. 'We've lost the AUV.'

At that moment, as I dragged myself from my bunk, I wondered if that was it. Was it all over before we had even started?

In the hours that followed, everything humanly possible was done to locate AUV 9. There was one major unknown: was it trapped beneath an ice floe, or had it escaped into the open ocean? If the latter, then we were in serious trouble. There was a 2.3-knot current running, everything was in motion, and there were floes as far as the eye could see in all directions – AUV 9 could be behind any of them, and we would never know. If, indeed, it was 'out there', somewhere between us and the white horizon, then our chances of finding it were in freefall.

There was, however, one thing that indicated it was most likely under the ice. AUV 9 is fitted with a beacon that communicates with the ops room via the Iridium satellite network, whose 66 satellites cover the entire earth including the polar regions. A property of the Iridium signal is that it does not pass through water and therefore, since we were not receiving a signal, the likelihood was that AUV 9 was trapped beneath an ice floe. Unfortunately there were ice floes everywhere, and although the AUV was quite likely to be tucked under the one beside us (where it had conducted the mission), we could not be certain.

While the technicians and data analysts were brainstorming, the scientists stood in the observation shelter above the bridge with binoculars, scanning the surrounds. At the same time, the fast rescue craft was searching the far side of the floe with drones airborne. Deploying the FRC in these conditions was not without risk and I could see that Captain Bengu, on the bridge, was tense with concern. The wind had risen to over 20 knots and although the FRC pilots were in radio contact, we could not always see them. They were, of course, wearing survival suits, but if something happened and one of them ended up in these freezing waters, they would not last long.

All morning the ROV team had been racing to re-ballast the ROV for a shallow under-ice dive to help in the search for AUV 9, and by midday they had it launched and on its way beneath the floe. But the floe was over four miles long and three miles wide and the tether would only allow it to explore a tiny area. On top of that, they were experiencing zero visibility. It was a dog-eared old floe of multiyear ice whose five to eight metres of thickness did not allow the passage of light. The darkness would not normally have been a problem as the ROV was fitted with powerful LEDs, but directly beneath the floe they had encountered what appeared to be a dense, algae-like bloom, so thick that it reflected the ROV's lights right back at the cameras. It could have passed within 20 inches of AUV 9 without spotting it.

The day ground by, with all of us in a state of utter frustration. Nothing worked. Tempers became frayed.

Our break came in the early evening, when the captain had to dampen down the ship's thrusters for a while. It was at that moment that our chief data processor, who had not left his post in the online room since the crisis began, picked up a ping. A single, feeble ping. It was AUV 9, struggling to communicate with us. That was a turning point: not only did the lone ping give us a rough direction of where AUV 9 was, it also confirmed that it was somewhere under the floe right beside us. More than that, it cut through our gloom like a sunbeam of light and gave us back something we had almost lost – hope.

NOON POSITION: 65° 48.069' S, 060° 32.890' W

17 JANUARY 2019

The continuing search for AUV 9

That single ping gave us hope we clung to stubbornly as the search continued.

However, the weather began to deteriorate and yesterday the wind was registering 35 knots and rising, with gusts coming in at over 43 knots. What was more, the floes by then were racing along at up to four knots. In such conditions it was impossible to dispatch drones and it was no longer safe to send out the fast rescue crafts.

The fear now was that AUV 9 might bump along the submerged surface of the floe and then emerge on the other side, only to be swept off by the current into the vast white unknown. If, however, it held its position under the ice, we would have to break our way through to it using the ship as a cudgel. Before we could begin that process we needed additional bearings on the AUV – and to achieve *that* we had to manoeuvre the ship along the edge of the floe, listening out for more pings.

But then, surreally, the ship's tannoy crackled to life as John Shears, our expedition leader, told us that if we wished to witness the remarkable sight of a massive iceberg slamming into a huge floe we should all go immediately to the viewing deck above the bridge. Everybody stampeded up there at once.

The berg drove into the floe with a low *crump* and stayed there. Watching it, we realized it was exactly where we needed to be in order to listen out for more pings from our stranded craft.

Eventually, with the thrusters on low, we secured enough pings to establish four bearings; where they crossed it gave a range, or distance to AUV 9, of 900 metres. This put it far beyond the reach of the ROV, which meant our only solution was to use the ship as a battering ram and smash a path through the ice. A red flag was placed on the floe at the position given by the analysts and, at 2253 hours, there was a *whoomph* and the whole ship shook as the *Agulhas II* assaulted the ice for the first time.

A long time was spent charging the floe, each time wresting from it huge chunks of ice which were then swept away on the current. Every couple of hours, the ship would pause to listen for further pings. Twice the red flag had to be repositioned, either because the initial data wasn't as accurate as we had thought, or because AUV 9 was on the move. While all this activity was taking place, the floe was being carried in a northerly direction.

By evening, everyone involved was in an advanced state of exhaustion and had to be stood down to sleep. Meanwhile the scientists, who until then had been observers of the evolving situation, were put out on the ice to conduct fieldwork.

NOON POSITION: 65° 46.292' S, 060° 31.240' W

18 JANUARY 2019

The search for AUV 9 continues

Today we have been chomping our way through multiyear ice, over four metres thick in places. There is not much that can withstand 60,000 tons bearing down upon it at seven knots. The vessel's great rounded stem rises up and over the ice, sometimes reaching more than 25 metres into the floe, and then, like a wrestler performing a flying body splash, comes slamming down with all its weight, crushing everything below and taking a bite out of the frozen slab up to 40 metres deep. The whole ship judders and there is a loud rasping sound as the broken chunks work their way along the side until they end up in our backwash. There is something so gloriously anarchic about it all.

All day yesterday the ship's engineers were crafting a special bracket for the HiPAP that would allow it to tilt 90 degrees in order to transmit directly with AUV 9. It was installed overnight and straight away we were able to determine the lost vehicle's position with almost pinpoint accuracy. Tactics now changed. With our improved positional data, we relocated the red flag and then began to gouge a huge V-shaped notch into the ice, aiming directly at the marker. By breakfast this morning we were 160 metres from our target. Everyone on the bridge, and in the lounge on the seventh floor looking out over the vessel's mighty bows, was brimming with optimism.

After breakfast the ship went alongside the ice with the intention of sending in the ROV to perform the extraction. However, ice conditions by then had deteriorated and soon it became too hazardous to deploy the ROV. The decision was taken for the ship to resume icebreaking, and by 1630 hours we were 50 metres away from the flag. The captain had adjusted his methods and we were nibbling our way forward with infinite care, so as not to inflict any damage on the AUV.

At 2230 hours, all icebreaking ceased. The red flag – which we could now see was not an actual flag, but somebody's T-shirt – was rippling in the wind just 20 metres from the edge of the floe. Ice conditions had improved again and the Eclipse team, who had been on standby all day, immediately launched their ROV. Within minutes, it disappeared under the floe. They were pleased to find there was much less algae in the water than before and the ice was much thinner, to the extent that the frozen ceiling above had a luminous quality with the light that was filtering through.

About 20 minutes into the dive, one of the ROV pilots, Steve March, tilted the main camera upwards – and there, silhouetted against the backlit floe, was the unmistakable torpedo-like form of AUV 9. By that time just about the whole ship was in the main lounge, where the ROV's video was playing in real time. The moment we saw the darkened cut-out of AUV 9, there was an eruption of undiluted joy that was heard by the cook in the galley three decks below.

'We have a visual on the AUV 9,' said co-pilot Dave O'Hara in a matter-of-fact tone of voice.

We watched in tense silence as the ROV struggled to loop a strap around AUV 9's tail fins. Finally, after 20 minutes of effort, it looked as if the rope might just hold long enough to draw the AUV back to the ice edge, where it could be secured by the FRC crew. Then – just as everything was looking so good! – a large block of ice slipped under our stern, threatening the ROV's umbilical at the point where it entered the water. The bridge was alerted and immediately the thrusters were activated so that the stern swung a few metres away from the rogue chunk; but this also moved the afterdeck away from the ice edge, which drew in on the umbilical. That, in turn and without warning, pulled the ROV backwards, allowing the AUV to slip free of the strap. It took a few moments for the pilots to realize what had occurred.

'I can't see it,' said Steve March, more to himself than to his boss, Steve Saint-Amour, who was managing all operations on the deck outside.

There was a bleak pause. Then Dave said, 'The fish has gone, mate. Repeat, the fish has gone.'

If the AUV had really disappeared, it could only mean that it was sinking. There was a further pause for thought before Steve Saint-Amour responded, 'Then we'll have to go chase it, goddammit.'

Everybody's eyes were once again pinned to the screen. The unthinkable had happened. AUVs are not meant to sink. AUVs are meant to be neutrally buoyant, or have a slight positive buoyancy; but of course, the day before its loss, AUV 9 was re-ballasted in order to give it the extra weight it needed to dive.

Now it was in slow freefall and the ROV was chasing after it. This was not easy on the submersible robot, as its ballast had only been optimized for a shallow dive. Twice the ROV's manipulating hand grabbed the AUV by its dorsal lifting lug – but the lug was small and the ROV could not achieve the purchase it required, and twice the AUV slipped away. All of this was conducted in slow motion as they were sinking. It had a dreamlike, almost balletic quality to it, like a pas de deux from a ghostly *Giselle*.

There was a little over 400 metres of depth at that point and when the pair reached 350 metres, it was decided to give up and allow the AUV to settle on the seabed. Once in a passive, stationary state, it would (we hoped) be simpler to recover with the winch. That was, until we realized that both the ROV and AUV were still under the floe. Throughout our endeavours the AUV, and indeed the ship, had been moving along with the ice, but now the AUV represented a fixed point on the seabed, while the mighty floe had just kept on rollin'. Icebreaking to get in over the AUV was not an option, and besides, we had to get the ROV out of there fast, as we were reaching the limit of how much umbilical we could unspool.

A hasty meeting followed, at which point it was decided to leave the AUV where it is so that the current can carry away the floe. We will try again in a few hours, when, hopefully, we will have open water – unless of course another floe has moved in to take the first one's place.

Though unwanted, this lull in the urgent activity does have its upside. Nobody can quite remember when they last slept.

NOON POSITION: 65° 48.113' S, 060° 40.517' W

19 JANUARY 2019

The recovery of AUV 9. Shackleton's South

Before breakfast today, Steve Saint-Amour rallied his team. It turned out that none of them had actually gone to bed because there had been too much to do to prepare for the next dive. There was open water over AUV 9 and by 0800 hours the ROV was on its way down. We had bright skies, perfect calm, and it was freezing cold.

For once, everything that followed was textbook. The descent to the seabed experienced no complications and soon the ROV team had a sling around AUV 9, after which they slowly raised it to the surface, where one of the small working boats took charge and towed it across to the ship where it was hooked by the crane. There are 95 people on board the *Agulhas II* and, although nobody was counting, I am sure every single one of us at that moment was hanging over the rails at the stern.

So what was it, exactly, that put us through all that torment? A software glitch, apparently – the one thing that was out of our control. Holly Ewart, a young Ocean Infinity project manager from London, summed it up best: 'It was high-end technology that put us in that mess, but it was human ingenuity, determination and professionalism at every level that got us out of it.'

She's right, of course, but a worrying amount of time has been lost, and everybody is anxious to resume the science programme without further pause. As soon as the submersible was back on board, the crane slung with cargo nets 'fished' two large chunks of ice out of the sea for engineering tests. The ship is now heading back to the Larsen C ice shelf, stopping along the way to perform further CTD scans and coring work.

I've been thinking about constipation! For us on the *Agulhas II* it is not a concern. We have a healthy diet, we are active and there is a gym. However, for Shackleton's men it was a serious problem. They had a diet of mainly seal blubber and dog food, and for weeks on end they could only loll about in sleeping bags all day.

We would not, however, know that they suffered from this condition, were it not for a single unexpected, throwaway line in one of the unpublished diaries: 'With our almost absolute meat diet and our lack of exercise, there naturally results chronic coprostasis with all its attendant evils.'

This inevitably makes you wonder about the bathroom arrangements for the 28 men. We know (also from the diaries) that at night they peed into a petrol can, but what of the other stuff? Again, a throwaway diary line provides a sort of answer by mentioning a skua feasting in what was, clearly, the latrines.

This is where the diaries serve a unique purpose: they are more truthful, less self-conscious and much richer in detail than *South*, Shackleton's published account of the expedition. That is not to say, however, that there are no problems with the diaries. There are, in fact, lots of problems, the main one being that the diary-writers knew they would be read by Shackleton, who was always banging away to them about the importance of loyalty. Indeed, in most cases if not all, they knew they would be obliged to hand their diaries over to him when he came to write the book everybody knew he was planning.

Naturally, this made the diarists very guarded indeed about any criticisms they might have had. However, in the event it seems as if only a few of the diaries were closely consulted for *South*. I can find a couple of extracts from Orde-Lees and Wordie, but that's about it; which is a pity, because some of the diarists were fairly deft with a pen, and often their insights and descriptive capabilities surpassed those of Shackleton himself. *South* would undoubtedly have been a better book if he'd made more use of them.

Shackleton did not write *South* himself, either; it was ghost-written. There can be little doubt that he was a literary man and certainly he had a love of the word and an orator's tongue to go with it, but he was not a man of the quill like Scott. He knew that he was sitting on a huge story, perhaps one of the greatest stories of survival ever, but he also knew that if it was not shared with the public it would disappear and might as well not have happened. He needed an amanuensis, somebody to cast his monument in print and, at the same time, produce a book that would help pay off his creditors.

He found that person in Edward Saunders, a journalist from New Zealand. The two had worked together before on *The Heart of Antarctica*, an account of the *Nimrod* expedition published to considerable acclaim in 1909; Saunders had actually written that book, but refused to accept any credit for it. Unsurprisingly, when the time came for Shackleton to write the story of the *Endurance* expedition, he again turned to his old friend. As before, Saunders wrote it down and shaped it up. Once more, as Shackleton had probably expected, he refused any acknowledgement whatsoever – and once more, thanks to him, the book was a critical success. Without any credit being given in the text to Saunders or his editor, the impression is that *South* was written entirely by Shackleton. Of the many omissions in the book, this is probably the greatest and most misleading.

NOON POSITION: 65° 49.076' S, 060° 35.683' W

20 JANUARY 2019

International Penguin Awareness Day.
'*Sir Ernest bid us come and slay penguins . . .*'

I have a confession to make. I have eaten penguin.

These days, I do not personally know of anybody living who has had that experience. I was born in the Falklands, though, the penguin capital of the world. As a young boy I spent my holidays on New Island, the most remote and westerly of all the inhabited fragments about the archipelago. Between there and Cape Horn there is nothing but miles of snarky, snarly, ship-gulping sea.

Life was basic on New Island and, although we never thought about it in these terms, it was all about survival. It was the view of my Uncle Cracks, who owned the island, that anything which ate grass, swam or laid eggs was only on this planet to be eaten – and that included the penguins, whose rookeries we could smell from the house. We were not the only ones in those days who ate them, as was made evident by an amateur penguin recipe book that did the rounds at the time, called *Fit for a FID*. Its appearance, sometime during the back end of the 1950s, upset my Aunt Agnes no end. The author had, apparently, appropriated one or more family recipes she had loaned to him at the Rose Hotel in Port Stanley, where the FIDS (young men with weak beards who worked for the Falkland Islands Dependencies Survey) stayed on their way through to Antarctica.

When I tell people I have eaten penguin, their first question is almost always 'What did it taste like?' To be honest, I don't really remember – I was very young. But what I do recall is Aunt Agnes announcing at breakfast that it would be penguin for lunch, in a way that suggested this was something we should be looking forward to as a treat. This brings me back to Shackleton, because the other day when I was reading Chippy McNish's diary for 11 April 1915, the following line (typically lumpy and unpunctuated) just jumped out at me: 'I had a walk along the young ice with McLeod for almost 4 miles it was a treat and good going had penguin for dinner it was a treat . . .'

If you were a penguin in the Weddell Sea in 1914, then Shackleton was decidedly bad news for you because you were very much on his menu. With 28 men to feed and twice as many dogs – well, that's a lot of dead penguins. And indeed that is how it was. Thousands of birds fell to what one diarist calls 'our murderous rapacity'. Again, from reading Shackleton's account in *South* you get no sense of the brutality by which the castaways lived – as usual, to get closer to the truth one has to dig within the diaries.

The main species of penguin encountered by Shackleton's men were emperors and Adélies. Then as now the emperors were concentrated mainly around the eastern and more central regions of the Weddell Sea; the Adélies were a bit more widely distributed but could be found in greater numbers around the Peninsula, where their breeding hotspots were mostly located. Thus the emperors receive more mentions at the beginning of *South* and in the early diary entries, but by the time the men have completed their drift the talk is mainly about Adélies. A diary entry from mid-December 1915 reads: 'The emperor penguins have been our stand by [i.e. when they could not get seals] ever since we came into the pack, even in midwinter when we were short of meat for the dogs we got them.'

For eating, emperors were probably the more favoured penguin because of the additional meat they carried. Adélies, by contrast, had much less to offer. The emperors were also easier to catch than the smaller and more nimble Adélies. Later, though, when they were on Elephant Island and seals were few, the men came to depend on Adélies and chinstrap penguins both for food and fuel.

As those of us on the Weddell Sea expedition have observed, emperor penguins are insatiably curious. When we are out on the ice they will walk over, rather stiffly with a measured tread and heads held high, to watch what is going on and, indeed, to examine us personally. I suspect they see us as just another species of penguin, but with sad beaks, terrible plumage and appalling posture.

On board the *Endurance*, they were often held in a pen specially constructed by Chippy McNish. Shackleton mentioned an emperor being quietly led back to the *Endurance* by one flipper, as if holding hands; once back at ship, it was corralled in an igloo to await butchery in the morning. During the early days, the bodies first went for scientific dissection. However, once there was nothing new to learn, the birds went straight to the galley.

While not everybody agreed with the slaughter of penguins, one diarist commented: 'Today a day of slaughter. The ice is dark with penguins. One flock of over 200 together, and many smaller ones on the ice. We got about 300 altogether . . . This apparently wholesale slaughter is most necessary. We are a big party, and this number of birds is less than a month's food.'

On Elephant Island the butchery of the Adélies was as relentless as it had been on the ice, but now there were also chinstraps and gentoos in significant numbers. By this time the stores salvaged from the *Endurance* had all but run out and penguins had become fundamental to the men's survival.

As when they were on the ice, the birds had to be gutted quickly before they froze. Normally this would not have been much of a challenge, but when Shackleton's men first arrived on Elephant Island they were all weak and suffering in varying degrees from frostbite.

On days of a major kill there was always a surplus of meat. This was stored in the snow, where it quickly froze. By the time they were rescued from the island by Shackleton, the men were, wrote Hurley, 'heartily sick of being compelled to kill every bird that comes ashore for food'. Others, notably Wild and Orde-Lees, felt the same.

We might wish to shield our eyes from the horrors of what Shackleton's men did to those birds a hundred years ago, but what the modern world has visited upon them is infinitely worse. The way in which we have blighted and befouled their environment has disrupted their feeding patterns and breeding cycles to the extent that many of their nesting grounds on the west side of the Peninsula are now ghost rookeries. Were Shackleton alive today, I have no doubt he would be a fervent environmentalist.

Last night we paused ship in front of the Larsen C rim to conduct further marine biological work; but not for long, because within an hour we could see a tight latticework of sea ice bumping its way along the shelf edge towards us at over two knots. By breakfast we were ensnared and had to recall the ROV. Once it was safely back on board we carefully threaded our way out of the tight ice field in which we had become enmeshed and headed away from the shelf in search of an open patch in which to conduct sea trials of AUV 7, an older and rather more bruised version of AUV 9.

An area of open water was found in which wet tests of AUV 7 were conducted. In general, they went well. This was particularly good news for me, as it's AUV 7 that will be leading the search for the *Endurance*.

NOON POSITION: 66° 23.422' S, 060° 19.081' W

22 JANUARY 2019

*Shackleton's dogs. Worsening ice conditions
around giant iceberg A-68*

I was talking to one of the American AUV techies the other day about our kids, home improvements and so forth, and I asked him if he missed his wife. He furrowed his brow, pursed his lips and arched his fingers, then extended his arms across the table towards me, spreading his hands and studying his cuticles. 'Nah,' he said at last. 'But I sure miss my dog.'

I think he was joking, but it got me to pondering the relationship between man and dog on the *Endurance*.

Learning from his thrust south with Scott in 1902, his own *Nimrod* expedition of 1907 and, of course, the 1911 triumph of Amundsen, Shackleton knew that the achievement of his goals depended upon dogs. Sources disagree on the number with which the expedition started, but it would have been between 68 and 70. All of them came from the Canadian Arctic, where they were bred for their strength and resilience.

Once the *Endurance* became trapped, Shackleton decided to move the dogs off the ship and into special quarters on the ice. This was much welcomed by the crew, for it not only got rid of the noise they made but also the smell of their excrement. At first they used the kennels from the ship to house the dogs, but soon they found it was better to make small igloos for them with blocks of ice cut from around the stern. Once the walls were up these were roofed with boards and frozen sealskins. Snow was then packed over the top and, finally, everything was cemented into place with a couple of buckets of water, which froze solid in seconds.

Although still tethered, the dogs seemed delighted with life on the ice, preferring even to sleep outside rather than use the shelters they had been given. During blizzards they would curl up into tight balls to retain heat and then, when the snow had built up in drifts over them, they would stand up, shake it all off, bark a bit and settle down to sleep once more.

Towards the end of January 1915 it was becoming evident that the physical condition of some of the animals was in decline. Six weeks later one of the more dominant dogs in the group, Saint, died from appendicitis.

On 14 August 1915, Orde-Lees wrote: 'Judge, Satan, Sandy, Sooty and Roy have all been mercifully despatched by Wild with his revolver.' It seems that the dogs' livers were riddled with redworm and the expeditionary team had forgotten, along with spare gramophone needles, to bring worm powder.

Throughout March and April the general condition of the dogs continued to deteriorate. On 24 March, two more had to be shot. Numbers, however, were raised when one of the dogs that was thought to be ill suddenly produced a litter of puppies.

The training and exercising of the dogs occupied most of the expedition's time. By the middle of March the dog teams, and indeed their handlers, had become highly efficient. Each team consisted of nine dogs, one of which was the leader. The choice of leader was often made by the dogs themselves, the dominant dog nearly always being the one that displayed most fighting spirit during the interminable squabbles that went on within the pack. Savagery and strength were always rewarded. Within the dog teams any skulking or disobedience was punished by the leader of the pack and, in general, the human driver only intervened when it was necessary to avoid a fatality. Every day, weather permitting, the drivers would take out their dogs and harness them for runs with weighted sledges. As for the drivers themselves, they could sit on the sledge if they wished, but because of the prevailing below-zero temperatures, they usually preferred to walk or jog along beside their team.

For Shackleton, the welfare of the dogs was a priority. It was decided to feed them as well as possible without touching their sledge rations, which the teams would need for the crossing. Around the middle of April the men began feeding the dogs pemmican, a fat and protein concentrate made mainly from dried ground meat, but the main part of their meals consisted of whatever wildlife was unlucky enough to come within sight of the ship. The area around the kennels became a grisly sight, strewn with lumps of blubber, seals' heads and half-eaten penguins.

We have 25 scientists on *Agulhas II*: the full petri dish, you might say. We love our scientists, honest we do, but we also love it when they get a poke in the eye. Today Captain Knowledge conducted an inspection of the ship and found that 'the science labs were dirty and needed to be cleaned'. Then came the really exquisite bit: just to be clear that the ship's domestic staff would not be doing it, he added that the ship would provide 'brooms, mops and a bucket as well as sponges, cloths and detergents'. I read these words aloud and they had an almost poetic ring.

John Shears and I were with the ice skipper, Captain Freddie, today. He is much concerned about a change in the ice dynamics. Floes and bergs are now spilling around the northern end of megaberg A-68 and piling into the area where we are currently situated. John has warned the scientists to prioritize their work in case we have to hightail it out of here.

NOON POSITION: 66° 25.890' S, 060° 10.088' W

23 JANUARY 2019

*The death of Shackleton's dogs. Attempt once more
to reach Larsen C ice shelf. Blocked by
fast-moving sea ice*

There were six dog teams with the *Endurance*, and their drivers considered themselves a cut above everybody else in the party. As for the dogs, they were a fierce pack of cross-bred, well-muscled, cold-hardened brutes that would kill if ever they were allowed to get out of control. According to Hurley, their successful training could only be achieved through 'frequent and judicious application of the whip . . . "Spare the whip and spoil the dog" is the imperative motto of all dog drivers.'

Macklin made frequent mention of his dogs. One of them, Mac, an obdurate bully, was more troublesome than most of the others, and Macklin once had to intervene in a bloody tussle between him and Paddy before the latter was seriously injured.

All the men knew that the dogs would have to die. When a decision had been made to remain at the site they called Patience Camp, the role of the dogs was reduced to bringing in the remains of slaughtered seals. They also had large appetites and were consuming food that was needed for the survival of the party. Nonetheless, when the time came for them to be destroyed, most of the handlers were grief-stricken. Wild, Shackleton's right-hand man (who had taken with him from the ship a revolver, a .33 calibre rifle and a 12-gauge shotgun) was willing to undertake the task in order to spare his friends the pain, but often the drivers felt it was their duty to perform the task. Macklin described his experience in particularly poignant terms:

I shot Sirius today with a shotgun. It went horribly against the grain to shoot this fine young animal, which I had brought up and trained from his birth. My hand trembled so that I could hardly do it, and he was all the time making joyous overtures to me. People at home could never understand what it meant to me, for these pups had been my daily companions in my walks down here where companions are scarce.

The slaughter of the dogs began on 14 January 1915: 54 days after the loss of the *Endurance*. When Shackleton informed Macklin that it was the turn of his dogs next, one by one they were unharnessed and then coaxed behind an ice ridge. Each was sat down and Wild placed the muzzle of the revolver against its head and pulled the trigger. Macklin, feeling almost sick, then gutted and skinned the animals in preparation for eating.

South tiptoes around the fate of the dog teams. This is understandable; the British, after all, are a nation of animal lovers and while it is okay, and even to be expected, for explorers to die, the same cannot be said of their dogs.

This morning we headed back to the Larsen C ice shelf to conduct a descent with the ROV down the face of its submerged keel. As a lifelong diver, I've been awaiting this experience with some enthusiasm – so I was dismayed when, before lunch, I went up on the bridge and saw a broad tide of sea ice, punctuated by a couple of large well-worn bergs, converging upon us at a couple of knots. Sure enough, soon after that Steve decided it would be unsafe to deploy the ROV.

Later we found an unclotted patch of water and the ROV went down on a biological sampling mission that lasted well into the evening. Shortly before midnight one of the marine biologists, Michelle Taylor, invited everyone to the wet lab to see the corals, anemones, starfish and sea squirts they had recovered.

The high point of the day for me, though, came when John Shears announced that the South African government has confirmed a four-day extension to our charter. Science work that was scheduled to end tomorrow will now continue until the 28th of this month, after which we will begin the search for the *Endurance*.

NOON POSITION: 66° 13.028' S, 060° 21.165' W

24 JANUARY 2019

Weird and wonderful. AUV 9 goes back under the ice.
Robertson Island. The collapse of the Larsen B ice shelf

We are told in the movie *Ghostbusters* that 'the purpose of science is to serve mankind'. I am not always so sure. We humans need our mysteries. They are there to enchant and delight, not always to be solved.

It was while pondering all this today that a painful little thought intruded upon me. Everybody who is not made of stone loves a good shipwreck, and the *Endurance* is the pre-eminent submerged tease of our times. There cannot be anybody who has heard the Shackleton story who has not, at one time or another, wondered what she looks like. Right now everybody, in their dreams, can make of her what they will; she is fuel to everyone's imagination. But if we find her then we will have revealed all, and perhaps I will be bursting bubbles.

Following AUV 9's successful open-water bathymetric survey, it was dispatched again today, this time on a short three-hour mission under a floe with its multibeam in upward-looking mode. There were some nerves about sending it back beneath the ice, as memories were still raw from when it was lost last time; however, all pre-dive checks were successful and at 0700 hours, the start key was 'pulled' and it was released down the chute and into the sea. It then dived to the loiter station, where its inertial navigation functions and other payload systems went through one final interrogation before Channing Thomas, the AUV team leader, affirmed his satisfaction. Then, once again, it was let loose under the ice in an unsupervised state.

As soon as the AUV was back on board, at 1100 hours, the ice scientists were lowered onto the same floe in the rope basket at the end of the crane to conduct snow analysis and drone radar surveys. During the last 36 hours there have been growing concerns over deteriorating ice conditions, which could, in tandem with persistent unfavourable winds, endanger the ship. As soon as the scientists were back on board at 1700, we left Larsen C and beat a retreat north.

Along the way we crossed the 100-mile gulf between the Jason Peninsula and Robertson Island. The view was mouth-watering, so I layered up and went out on deck. I was watching a pod of minke whales and thinking how utterly beautiful it all was when, as if out of nowhere, a dreadful truth dawned upon me. It was the realization that the beauty of everything I could see was false. Nothing upon my horizon was as nature had intended.

What I should have been seeing were soaring ice cliffs, tens of thousands of years old, the chiselled face of a wall-to-wall glacial shield that ran for miles and was part of the great weave essential for the health and habitability of our planet. This was where the Larsen B ice shelf should have been.

Just over 15 years ago, we – humans, that is – by our reckless stewardship of the environment, reduced this fundamental, life-driving force of nature to nothing. In the space of less than three weeks all 1,250 square miles of it had gone, as warming temperatures caused it to collapse into the sea and disintegrate.

NOON POSITION: 66° 04.387' S, 060° 35.542' W

25 JANUARY 2019

The construction of the Endurance. *Larsen Inlet.*
AUV 7 deployed on hydrographic survey mission

The *Endurance* is at the very heart of the British cult of Shackleton, but the *Endurance* is not British. She was conceived by a Belgian and built in Norway.

Adrien de Gerlache was a Belgian polar explorer of renown who conceived the idea of building a comfortably appointed, icegoing steam 'yacht' with 10 cabins to carry rich tourists to the Arctic to shoot polar bears. With this as his business plan he went into partnership with Lars Christensen, a young Norwegian ship-owner, who would oversee the construction of the vessel. Work began early in 1911 at the Framnæs shipyard in Sandefjord, Norway, and was overseen by the well-known ice ship builder Johan Jakobsen.

The 144-foot-long *Polaris* was launched on 17 December 1913, but by that time de Gerlache had experienced some financial difficulties and the *Polaris* was immediately put up for sale. Fortunately for Christensen, it was at this point that Shackleton came along, and though the *Polaris* was not exactly what he was looking for, she was adequate. A knockdown price of £11,600 was agreed. Shackleton changed her name to *Endurance* (after his family motto, *fortitudine vincimus* – 'by endurance we conquer') and in the spring of 1914 she arrived at Millwall Docks, London, to begin refitting for the Antarctic.

As a maritime archaeologist, I am hungry to know every detail there is to know about any ship I am investigating: how she was built, the political environment of the day, the economy within which she operated, the purpose of the voyage and her cargo, and, of course, the people and society on board. And, when possible,

I like to stand on the very spot where the vessel was made and just commune with history.

A little before Christmas 2018, my wife and I went to Sandefjord to see where Jakobsen had built the *Endurance*. A road now runs around the edge of the fjord but at one point it bends to accommodate a cleft in the wooded hillside behind, and it is here that the best timber-built ice ships in the world were once fashioned. The lofts, carpenters' shops and sawmills have long since gone, and the waterfront is now dominated by a marina for small boats. As for the spot where the stocks once stood, there is now a small parking area with a boulder at one end which, at first glance, looks a little out of place. Take a closer look, however, and you will see a plaque on the boulder which features a profile drawing of an elegant barquentine-rigged steam vessel, above which is written 'Polar wooden screw yacht *Polaris*', and then below, in capitals: 'ENDURANCE'. There follows a brief history of the ship in Norwegian along with a picture of the location taken in 1913, which shows the *Endurance* (or *Polaris*) during construction.

Under bruised skies, with the fjord in front and the Norwegian wood behind, we just stood in silence and thought about that little mistake of a ship which, from this Viking inlet, set sail on a voyage that would take her right off the map . . . and from there into legend.

Larsen Inlet is situated to the north of Robertson Island, between the Sobral Peninsula and Cape Longing, probably less than 50 nautical miles from where we were yesterday.

In normal conditions we could have reached the inlet in several hours, but the ice fields we had to pass through were dense and stubborn and a thick blanket of fog made it impossible to pick out the best leads (channels of open water) in advance. At times we were just crawling along at two knots. We reached the waters off the inlet around 1400 hours, by which time the fog had lifted and the ice had loosened up considerably. Nonetheless, it was not until late afternoon that we were able to launch AUV 7 on a 12-hour bathymetric mission.

There had been whispers that we'd left Larsen C too early, thus compromising the science, but now everybody could see it had been the correct decision. The Weddell Sea is a spit-in-your-eye kind of place: it can quickly turn nasty and its ice can have you by the throat in less time than it took the *Endurance* to complete her plunge. All it takes is a compacted surface, unfavourable currents, a drop in temperature and some persistently contrary winds. Throw in a blizzard and, if you are gashed and sinking, nobody can reach you.

NOON POSITION: 64° 46.492' S, 059° 11.433' W

26 JANUARY 2019

Frank Worsley, master navigator. Snow Hill Island.
Scientists winched out to take ice core samples

The quest for the *Endurance* depends entirely upon a single set of particulars that I cannot even regard as fact. It is the sextant position taken by Frank Worsley on 21 November 1916 – the day after the ship he captained slid beneath the ice and plunged to the seabed 3,000 metres below. Our search for Shackleton's ship has required a crack team of specialists, the most advanced subsea search-and-survey technology in the world and, of course, an icebreaker of exceptional capability, all of which has cost many millions. Yet it all hangs on these coordinates:

68° 39' 30" S, 52° 26' 30" W

Obviously, my initial step when considering the search had to be an evaluation of these coordinates. This came down to two considerations, the first of which was Worsley himself, who has passed into history as a brilliant navigator. My second consideration was the coordinates themselves; by which I mean the circumstances in which they were taken and the uncertainties they reflect.

When Worsley joined Shackleton's Imperial Trans-Antarctic Expedition, he was a 42-year-old mariner from New Zealand who had been plying the oceans of the world since he was 16. His career began in 1888 as a 'brassbounder', or junior midshipman, on the Tyne-built *Wairoa*, a full-rigged Cape Horn wool clipper of the New Zealand Shipping Company. It was on this ship, during a passage to Britain and back, that he learned the 240 ropes, felt the swish of a rope's end and, more importantly, began his study of navigation, something that would serve him well in the Antarctic

27 years later. Interestingly, just two years after Worsley joined the *Wairoa*, Shackleton, then also a teenager of 16, went to sea before the mast on a Cape Horn voyage – but, it seems, his experience of navigation was very different from that of Worsley.

The teaching of navigation to the apprentices in the British merchant service was overseen by the ship's master, the majority of whom, it was said (and Shackleton's was probably one of them), could not be bothered and offloaded the responsibility onto the mate, who probably shared the master's indifference and, because he was a product of the very same system, was likely to have only basic knowledge and motivational skills. Reading Shackleton's two books, one gets the impression that navigation was no more than a chore for him, a perfunctory set of procedures that he had to follow to know where he was. He knew, for instance, how to 'bring down the sun' to obtain his zenith distance and then add declination to arrive at his noonday latitude, but that was where it ended; he had no interest in navigation for its own sake. For Worsley, by contrast, navigation was a lifelong passion which can probably be traced back to his first captains of the New Zealand Shipping Company.

In 1895, seven years after he first went to sea, Worsley left sail for steam when an opportunity arose to become second mate on the SS *Tutanekai*, an old tramp that served the Pacific Islands. By then he would have been proficient in celestial position-fixing, but that alone did not make him a good navigator – for in those days navigation was part science and part art, and it was mastery of the art that made for excellence.

It's not until 1900, when he is examined for his 'foreign master's' certificate, that the historical record gives an indication of Worsley having become a mariner with more than an ordinary aptitude for navigation. In that year he was one of two candidates who sailed through their papers with distinction. His principal examiner wrote: 'He passed very creditably at first attempt, all problems being correctly worked without having been returned for any correction.' In a report on that year's candidates he singled out Worsley as 'deserving special mention for the very creditable manner in which [he] passed [his] exams'. But it was in a personal note from the same examiner that I found a line that further bolstered my confidence in Worsley's coordinates for the sinking of the *Endurance*. In part it read: 'I am glad to hear that you are keeping your hand in with the most useful problems in navigation.' This makes it clear that Worsley was not just adept with the sextant, but that position-fixing had become an enthusiasm for him at a more esoteric level.

Having gained his master's ticket, Worsley went on to his first command, the *Countess of Ranfurly*: an auxiliary-powered, three-masted schooner that was engaged in the South Sea Island trades. Making landfall between such small dots on the map over vast distances would certainly have honed his navigational skills.

From one ship to the next, Worsley continued at sea, becoming, along the way, a Royal Navy Reservist. In 1906 he left New Zealand and the South Seas for the North Atlantic, and it was during this period, when seeking training in the military aspects of seafaring, that he attended the Nautical College at Pangbourne and thereafter served a year on HMS *Swiftsure*. On returning to the merchant service Worsley was obliged to obtain an official letter of reference from the New Zealand government regarding his eight years at sea in their service. In part it read: 'The Prime Minister of New Zealand speaks of Lt Worsley [his then rank in the reserves] as a first class seaman and navigator.'

When, in 1914, Worsley famously poked his nose into Shackleton's office at 4 New Burlington Street and came out a few minutes later as captain of the *Endurance*, he had been 26 years at sea and had just served as second officer on one of the transatlantic freighters of the Allan Line Royal Mail Steamers. He had also by then been promoted to lieutenant-commander in the Royal Naval Reserve. From my scrutiny of Worsley's professional life before he joined Shackleton's team, I have no doubt that he had become a highly able mariner and navigator; but during Shackleton's expedition, it was only after they left the ice in April 1916 that his exceptional abilities as a navigator became evident. At that time they were heading in a westerly direction for the Bransfield Strait with the idea of heading down the South Shetland Chain to Deception Island, where there was shelter and possibly food.

On 12 April, three days after they had taken to their boats, Worsley managed to take a position. Until then, they all thought they had progressed 30 miles in the direction of the South Shetlands. Worsley's fix told him that they had indeed travelled 30 miles – but in the opposite direction. The current from the Bransfield Strait had been stronger than they thought. Everybody, even Worsley, doubted his position. By then it was obvious that Deception Island was unreachable. For a while Shackleton considered Hope Bay at the very tip of the Antarctic Peninsula, but when they met with long streams of ice they decided instead to head for Elephant Island, which was at the very top of the Shetland Chain.

In shaping this new course, Worsley's problem was that he was not certain of his calculated position. In navigation at that time, you had to first know where you were before you could determine where you were going. Worsley later said that 'the next three days were the most terribly anxious ones of the whole enterprise' because 'my "sight" had been taken in extraordinarily difficult conditions, being merely glimpses of the sun between icebergs on a misty day; and had my calculations been wrong in any way, it would have meant that twenty-eight men would have missed the land and would have sailed out to practically certain death'.

And yet, despite all the challenges, what happened next showed that he must have been correct in his original sextant determinations.

Under Worsley's guidance, the three open boats made Elephant Island. It was a 50-hour ordeal in which, frostbitten, seasick and numbed to the marrow, they rowed under a tiny sail across the implacable circumpolar current. Often the gale-force winds whipped up sheets of drenching spray which then froze over both the men and their woodwork. They took turns to chip away the ice as its increasing weight began to press down upon the sides of their boats above the water. Yet, through it all, they clung to Worsley's bearings, even while zig-zagging through patches of loose pack.

And then, on the morning of 15 April, in Worsley's words, after 'two days of . . . working in and out amongst the floes with no accurate means of checking compass courses and two nights drifting at the mercy of the winds and currents . . . we saw to the northeast the lofty snow-clad peak of Clarence Island. A little later the peaks and ice uplands of Elephant Island showed cold and gloomy, 35 miles to the north-northwest. *They were both exactly on the bearings I had laid off* [my italics].

Of his achievement Shackleton wrote: 'In the full daylight Elephant Island showed cold and severe to the north-northwest. The island was on the bearings that Worsley had laid down, and I congratulated him on the accuracy of his navigation under difficult circumstances, with two days dead reckoning while following a devious course . . .'

This was truly a remarkable feat of nautical wayfinding. Even Worsley appears to have been a little surprised by his precision and wrote rather modestly: 'To tell the truth there had been a large element of luck in making this landfall. I had been very anxious, for our lives depended on reaching land speedily.'

Aware that if they remained on Elephant Island they would almost certainly all perish, Shackleton decided to set out in the *James Caird* to seek help, despite the fact that, in his words, 'the ocean south of Cape Horn in the middle of May is known to be the most tempestuous storm-swept area of water in the world.' To accompany him he selected Worsley, the ever-dependable Tom Crean, the carpenter Chippy McNish and the seamen John Vincent and Tim McCarthy.

Port Stanley in the Falklands was closer at 540 miles, but it would have meant not ceding an inch of leeway to the circumpolar current while at the same time squaring north across the face of the prevailing westerlies with only a scrap of sail for muscle. South Georgia, at 800 miles, was much further away, but both current and wind would be pushing them in roughly the right direction.

For navigation Worsley had with him at least two compasses, a German blueprint map of South Georgia, his navigation books,

an almanac, and a Heath & Co. Hezzanith sextant that belonged
to the navigation officer on the *Endurance*, which Worsley must
have judged superior to his own. His chronometer for the voyage
was 'an excellent one of Smith's, the sole survivor, in good going
order, of the twenty-four with which we set out in the *Endurance* ...
I carried [it] slung around my neck by lampwick, inside my sweater,
to keep it warm.'

The crossing to South Georgia took 14 days, during which they
often battled gale-lashed seas that streaked the water white with foam
and sent the spindrift flying. At the height of these storms, towering
Cape Horn combers would come boring down upon them. At times
they rode the crests, with the water falling away steeply on either
side – and then, seconds later, they were swooping down into dark
hollows 40 or 50 feet below the summits of the waves, fearful that
at any moment they might be taken in a gulp. They worked in four-
hour shifts during which, according to Shackleton, 'one man had the
tiller-ropes, the second man attended to the sail, and the third baled
for all he was worth . . . A thousand times it appeared as though the
James Caird must be engulfed; but the boat lived.'

Typical of these parts in winter, the skies were almost contin-
uously overcast, making celestial navigation by day almost
impossible. At night it was worse. During the long hours of darkness, Worsley wrote, 'I steered by watching the angle at which the
[masthead] pennant blew out; at times verifying the course by a
glimpse of a star through a rift in the clouds.' His favourite star
was Antares, the red eye of the scorpion. Once in a while they
would light a candle stub for a few minutes 'to enable the helmsman
to correct the course and check it off by the sea, wind and fluttering
pennant.'

This was navigation by what is called 'dead reckoning', the
seaman's estimation of their direction and distance covered which
tells them where they *ought* to be, whereas, by contrast, celestial
sextant navigation tells them where they *are*. Worsley later gave a
dramatic account of what it was like trying to capture their latitude
from the *Caird*:

Once, perhaps twice, a week the sun smiled a sudden wintry flicker, through storm-torn clouds. If ready for it, and smart, I caught it. The procedure was: I peered out from our burrow – precious sextant cuddled under my chest to prevent seas falling on it. Sir Ernest stood by under the canvas with the chronometer, pencil and book. I shouted, 'Stand by,' and knelt on the thwart – two men holding me up on either side. I brought the sun down to where the horizon ought to be and as the boat leaped frantically upward on the crest of a wave, snapped a good guess at the altitude and yelled 'Stop'. Sir Ernest took the time, and I worked out the result . . . Since leaving Elephant Island I had only been able to get the sun four times, two of these being mere snaps or guesses through slight rifts in the clouds.

The whaling stations on South Georgia are all located on the northeast or upper coast, so ideally, coming from the southwest, Worsley's navigational target was the northwest tip of the island, from where they would make their way along the upper coast to the stations. If, however, he missed the end of the island and passed too far to the west, they would not be able to retrace their route because of the opposing winds and current, in which eventuality they would certainly all die. So on their 13th day out from Elephant Island, when they knew they were nearing their destination, it became essential that they establish their precise whereabouts by sun sightings rather than dead reckoning.

Two observations are necessary to fix your position, but during the morning when he observed for longitude, the boat was 'jumping like a flea' and the skyline was partially obscured by mist. If you cannot see your horizon it is impossible to measure the sun's altitude accurately, so naturally Worsley was mistrustful of his result. Then, at noon, when he hoped to establish his latitude, the sun was no more than a befogged blur and because there was no 'limb' (i.e. visible edge to the sun) he had to focus on its centre by guesswork. Ten times he performed the exercise and then took the mean to calculate the sun's altitude. When Worsley told Shackleton that he could not be certain of their location to within 10 miles, the Boss, ever cautious, decided that he could not risk running for the tip of South Georgia in case they overshot – so Worsley shaped their course to the east, in order to make landfall along the uninhabited underside of the island near King Haakon Sound.

Worsley wrote, 'My navigation had been, perforce, so extraordinarily crude that a good landfall could hardly be looked for' – but at one o'clock the next afternoon: 'McCarthy raised the cheerful cry, "Land Ho!" There right ahead, through a rift in the flying scud, our glad but salt-blurred eyes saw . . . Cape Demidov, the northern headland of King Haakon Sound.' Once again, Worsley had been spot on.

Shackleton once wrote that he 'had a very high opinion of [Worsley's] accuracy and quickness as a navigator'; and based on my review of his plotwork after they left the ice, so do I. Regarding his famous position for the sinking of the *Endurance*, I know that the sextant on that day could not have been in better hands. But even so, the position he recorded on Sunday, 21 November 1915, does reflect a number of uncertainties, meaning that our upcoming search for the *Endurance* will involve a rather discomforting element of speculation.

The survival of the Imperial Trans-Antarctic party has always been attributed to Shackleton's leadership, and certainly I would not wish to detract from that, but once they had left the ice there was a significant shift of responsibility. By his own admission Shackleton knew nothing of small boat sailing, whereas Worsley, from his time as a captain under sail in the South Seas, was an expert. It is doubtful that Shackleton's navigational skills would have been good enough for the combination of extreme dead reckoning and sharp sextant management that got them across those brutal seas to Elephant Island and South Georgia beyond. Studying Worsley as a navigator has left me with the abiding thought that once the men left the ice, the lion's share of the credit for their survival really belongs to him.

By 0430 hours today AUV 7 was back on board, having completed its hydrographical survey of the seabed off Larsen Inlet. As soon as it was secure within its cradle we began our transit to Bank B, a site off Snow Hill where we arrived in the early evening. Here AUV 9 conducted a final under-ice, multibeam sonar survey while the scientists were winched over the side and onto the same floe using the ship's crane.

Their objective was to conduct three last studies on the composition of the ice. The first, led by Wolfgang Rack from Canterbury University in New Zealand, involved the excavation of a pit to study the stratigraphy – the depths and densities of the snow and ice layers – in order to understand better how these factors influence the buoyancy of the floe. The second exercise, under the supervision of Annie Bekker from Stellenbosch University, required the extraction of core samples from the ice in order to determine its chemical composition, micro-crystalline structure, biological-nutritional content and, finally, its ability to absorb CO_2 from the atmosphere. The third scientific project, also led by Annie, was to use the ship to whack off a few large chunks from the floe which were then fished on board using cargo nets slung from the crane. From these she and her colleagues cut oblong samples, which they then 'bent', or put under weighted pressure, in order to establish their breaking strain and other mechanical properties. The accrued data from this and her team's other scientific activities will be applied in the design and manufacture of the next generation of ice ships.

There are bergs everywhere. Penguins and seals are relatively few and distant, but their appearances are increasing. Giant petrels, snow petrels, Cape petrels, sheathbills are all back in number, and some albatrosses too. A crowded field of huge tabular icebergs was particularly spectacular today. Most appear to be grounded, which you can tell from the horizontal stain marks left along their sides by the rise and fall of the tide. About 90 per cent of a berg normally sits beneath the surface, forming a huge keel, and when they are not bogged down in the seabed they can drift as fast as one metre a second, or up to 40 kilometres a day. I have spent much time on the bridge watching the ice in recent days and sometimes, particularly when the ship is stationary, you see huge bergs propelled along in one direction by the action of the current upon their keels while the surface pack, driven by the wind, is moving in another.

Today we passed a few miles to the west of where Shackleton and his men were drifting on 27 February 1916. On this day one of the diarists wrote: 'Over 100 bergs can be discerned with binoculars. Worsley claims to have counted 160 and he is not far in excess.' With binoculars I made 183, but I am sure if I had better eyesight and had been a little more attentive I could have raised that to over 200.

We are also back into short periods of night, which means we have witnessed some spectacular sunsets. Shackleton's men were awestruck by the evening displays they observed. One diarist wrote, 'We get beautiful sunset and sunrise effects, the snow flushes a beautiful pale rose colour, the tinting and lighting is delicate in the extreme. There is nothing harsh about Antarctic colouring at any time . . .'

Last night everybody was out on deck for one of the most emotionally squeezy sunsets I have ever witnessed. There was about it something so otherly and beyond that you wondered if you were really supposed to be here. It seemed almost as if we had trespassed into some polar hidey-hole where the gods go to drain their rainbows. Blends, spills and blushes were all draped with cloud and punctuated by the silhouettes of bergs thousands of years old. Nobody said a thing. I looked at the faces. At moments like this you often think of family, life and stuff that matters. My thoughts, however, were directed more to the sunrise and what it will bring – for tomorrow, you see, is a big day. Tomorrow the science ends and the search begins.

NOON POSITION: 64° 39.065' S, 057° 04.349' W

27 JANUARY 2019

*Science programme over. Leave Erebus and Terror Gulf and
begin transit to* Endurance *search area. Pass Joinville Island
at the tip of the Peninsula*

We are on our way to find the *Endurance*. At 0306 hours we officially
paused the science and set off on our long transit to where she
sank.

We are a joyous bag of wonks on this ship; we come from
every nook and rabbit hole on the planet, our T-shirts are shouty
and we wear every colour of sock; yet we are all card-carrying
Shackleton fanatics. It is this last quality that has led to what I
sense to be a quiver of optimism and expectation running through
Agulhas II.

This is the most unreachable wreck the world has ever known;
everything is going to be difficult. If it was easy, somebody would
already have found it. Although it is hard not to be carried along
on all the optimism, I am going into it with eyes wide open.

In broad terms, I see four major hurdles ahead of us, each a
level higher than the one before:

1. Can we reach the search area? We have over 80 miles of dense
   ice to plough through before we even begin the hunt.
2. If we can get there, can we launch and recover our AUVs?
   We need some open water for this, and right now there is
   ten-tenths pack ice coverage over the search area.
3. Will our equipment perform in these extreme conditions?
   We have the most sophisticated unmanned search submersibles in the world, but never before have they been stretched
   as they will be once they are under the pack.
4. Is the wreck actually within our search box?

This project is going to be a severe test of the ship, the equipment she carries and, indeed, us as people.

Before breakfast I was on the bridge with Captain Knowledge and John when, in glorious wide open skies, we passed Snow Hill and Seymour Island (both named by James Clark Ross during his Antarctic expedition of 1839–43). The captain pointed to the screen on the console, which had flashed up a warning to say we were in Argentine waters without authorization. Argentina has a base on the high grounds of Seymour, which I could just detect as an orange smudge through the binoculars. I asked the mate if they were going to respond.

'No,' he said. 'Everybody just ignores it. This is the only country that makes claims like that – nobody owns Antarctica.'

We continued our transit, and by 2100 hours we were off Paulet Island (approximately one mile in diameter), which is easy to identify because of its distinctive cinder cone summit and dark, snow-streaked mantle.

Shackleton had been eager to reach Paulet because he knew about the hut and food depot left there by the Nordenskjöld expedition. If they could have made it, the plan was to winter in the stone-built shelter and survive on the stores. Shackleton knew there were stores on the island because in 1903 he had assisted with the equipping and supply of the relief ship sent to rescue the Swedish party, whose ship, the *Antarctic*, had been caught in the ice and lost in the very area where we were today, about 25 miles from Paulet. The irony was not lost on Shackleton, who later wrote: 'We remarked amongst ourselves what a strange turn of fate it would be if the very cases of provisions which I had ordered and sent out so many years before were now to support us during the coming winter. But this was not to be.'

The problem for Shackleton was that the ice was no longer solid and interlocking but rather, since it was the very end of summer, a loose, jostling admixture of floes and brash, all melded into one by a thick, viscous, porridge-like slush. Although they passed 60 miles to the east of the island, they were unable to make landfall.

The *Endurance* lies somewhere beneath the massive gut of consolidated ice that characterizes the west central Weddell Sea at this time of year. It would take too long to punch our way directly through to the site; besides, we doubt we could do it, and, if we failed there would be an uncomfortably high probability of becoming beset. This would be very serious indeed, as there is no other ship in the region with the heft and strength of hull to come to our aid. There is, however, on the opposite side of the pack from where we are now, a large area of ice riddled with leads, which Captain Freddie and I have been watching closely via satellite for several days. We call it 'the patch'. Remembering that the Weddell Sea is one vast, clockwise-moving gyre, that patch is slowly edging towards the area where I believe the *Endurance* to be. It will not be over the site by the time we get there, but it will be close enough to shorten our final approach into the search area.

So we are now heading for the patch; but to get there, we first have to circle halfway around the outside of the pack, a distance of 420 nautical miles, which, at an average speed of seven to eight knots in relatively open water, should get us there in 52 hours. Once we have passed through the patch, the real battle will commence as we slam, bludgeon and claw our way northwest towards the spot where the *Endurance* went under. It represents a distance of only 75–80 miles, but it is going to be a brutal slugfest every inch of the way.

By late this evening the *Agulhas* was heading eastward across the top of the pack. Tomorrow we will wheel around onto a southerly heading. Every day I take out the triangles, parallel rulers and dividers and draw our course upon an Admiralty chart of the Weddell Sea. The route we have followed looks utterly erratic and whimsical. Anybody not familiar with these waters might think we've had a drunk at the helm, but every swerve, hard right and handbrake turn we have made has been in response to the ice.

As Shackleton once said when they were in their boats and picking their way through densely noddled waters, 'The old adage about a short cut being the longest way round is often as true in the Antarctic as it is in the peaceful countryside.' I heard our mate put it more succinctly when one day he growled: 'Down here we don't do bloody beelines.'

NOON POSITION: 64° 11.058' S, 055° 50.859' W

28 JANUARY 2019

Ice conditions over the search box. Introducing James Frank Hurley. Heading east across the top of the Weddell Sea pack

This evening it was my turn to give an after-dinner talk about my work.

For 20 minutes I talked about some of my previous excavations before we got to the real meat of the evening: a screening of Frank Hurley's film *South*, which I had brought with me from Oxford.

Hurley's moment came in 1911, when Douglas Mawson invited him to become the official 'camera artist' for his expedition to survey the sector of the Antarctic Continent immediately south of Australia. Three years later, Shackleton saw some of Hurley's work for Mawson and knew he had to engage the 29-year-old Aussie to immortalize his own grand undertaking.

Eventually, Hurley became much more than the team photographer; he was the one to whom everybody turned for help with anything technical. Shackleton called him 'our handy man'. He was a complicated character who could be a bit prickly – Macklin described him as an 'ungrownup Australian' who always 'had to be patted on the back', and even Shackleton was cautious when in his company.

But, of course, it is by the quality of his work that Hurley is now judged.

What is not generally known is that Hurley came close to losing all of his moving footage and many of his stills when, in November 1915, the ship slumped in the ice and his studio and its storage facilities became submerged. On 2 November he recorded in his diary how he managed to save most of his work: 'During the day, I hacked through the thick walls of the refrigerator to retrieve the negatives stored therein. They were located beneath 4 feet of mushy ice and, by stripping to the waist and diving under, I hauled them out. Fortunately, they are soldered up in double tin linings, so I am hopeful they may not have suffered by their submersion.'

Because of their bulk and weight, Hurley and Shackleton had to discard most of the photographic plates. Though unlikely, it is possible that some of those abandoned images are still on the site and may one day be saved.

Earlier I was looking at recent satellite imagery of our search area with Captain Freddie, the ice skipper. He, like all of us, is concerned about the thick, ten-tenths ice coverage currently smothering the site. As it is at present, even if we can bludgeon our way through to the *Endurance* search area we will not be able to launch and recover our AUV, or even lower our acoustic positioning system into the water beneath our keel. As Freddie said, 'What we need is some stiff weather that will get in there, create havoc and then be gone.' I am praying for storms again.

By breakfast we were heading east across the top of the pack, but it has been a slow day. At one point our speed was down to four knots as we dodged our way along in dense fog through what the mate called 'some bloody bergy stuff'. Everybody, however, is in a buoyant mood because after a month at sea, we are at last on our way to what one of Shackleton's diarists called the 'forsaken spot' to begin our search.

NOON POSITION: 63° 57.100' S, 051° 38.698' W

29 JANUARY 2019

The Endurance *as a battering ram. The team is briefed
on the search. East of the pack, looking for open water
in which to deep-test the ROV*

In recent days, I've been spending almost all my time on the bridge or in the viewing gallery above; I don't want to miss a single minute, as I know I will never experience anything like this again. Occasionally I watch Captain Knowledge, a quietly authoritative man of 39 from Durban, who, despite all the strain, I just know is enjoying it all. How can you not be when you are riding all this weight, size and power? At one point yesterday, as we were mangling our way through a field of particularly obdurate ice, he put down his binoculars, gave me a lopsided smile and said, 'This is all good practice for reaching the *Endurance*.'.

Today we encountered a couple of orbital ice fields that were far longer than they were wide. Rather than skirt their peripheries it was quicker to plough through them. Although their margins consisted of first-year ice that we barnstormed through with ease, the central parts were old, thick, gnarled and rifted – what the mate called 'some pretty badass stuff'. In these core areas we resumed icebreaking, trying to open up leads; when this was not possible we just crushed our way through, slamming, withdrawing, then charging again.

The *Endurance* was of course a very different type of ship and when she rammed the ice her technique was quite unlike ours. Whereas the *Agulhas II* drives up and over then crashes down, the *Endurance* used her sharp metal-sheathed bow to slice into the ice and create a crack, which she would then try to wedge apart in order to open a lead. Also, the *Endurance* could only tackle first-year ice of no more than three-foot thickness. In fact, when one looks closely at the diaries one finds that the *Endurance* did not actually do very much ramming, instead relying more on her bijou dimensions and sensitive steering to inveigle her way through the pack.

It has been another slow day. Overnight we were dodging along in ice-strewn waters in light fog at seven knots, but by breakfast the fog had thickened and speed was reduced to an even less dashing five knots. There are many large bergs and several ice fields blocking our way. We are now east of the main Weddell Sea pack and working south. We are hoping to find an area of open water in 3,000 metres (the depth of the *Endurance*) where we can conduct deep-water tests of the ROV. We have a location in mind which, according to satellite surveillance, looks pretty clear, and as long as conditions do not worsen we should be there by the second half of the evening.

At 1100 hours we held an all-expedition meeting in the ship's auditorium during which John, our team doctor Claire Grogan and I briefed everyone on our upcoming procedures, methodologies and objectives. The scientists are all busy writing up their reports while everybody else is resting in preparation for some sleepless days ahead. In a little more than 24 hours we will be at the ice edge, and then things will become intense. I have been reading some of the press reports; one made the interesting observation that we are heading to a spot where no ship has been since the *Endurance*.

NOON POSITION: 65° 25.712' S, 047° 54.295' W

30 JANUARY 2019

Disaster strikes

Sometimes you just want to throw back your head and howl at the heavens.

For a couple of days we've been looking for a clear patch in deep water where we can stress-test the ROV. Shortly before midnight last night, we came upon a suitable spot and hove to. There was no ice anywhere, the sea was calm and we had 3,000 metres beneath our keel. The Eclipse team had already gone through their pre-dive checks and the ROV was poised at the stern, ready for launch. The objective was simply to descend to 3,000 metres, perform some checks and then head back. It was so routine, nobody gave it a thought.

The ROV submerged at 0248 hours. Everything proceeded without a hitch until it was at 2,976 metres – just 24 metres from our intended test depth, and well over 1,000 metres above what we call 'crush depth'. The clock was on 0533 hours when suddenly we lost telemetry and the screens went black. They didn't even flicker, as usually happens when there's a problem – from one instant to the next, they just went black. All of them.

Steve March, who was co-piloting, said he at first thought it was just 'something to do with the fibre connectors, maybe a bit of water penetration. The usual response is to take it up a hundred metres to relieve the squeeze and, if everything is working again, you just proceed as before, but maybe a bit slower. So we took it up, and still there was no response from anything so at that point the boss gave the order to abort.'

It took two hours for the ROV to return to the surface, at which point the focus was on recovery. It was not until it was secured on deck that the inspection began. The first person to take a look was its pilot, Dave O'Hara, a burly guy from Belfast who, when not on ROV ops at sea, organizes music festivals. The front of the ROV looked all right so he went round to the port side; there he paused before bending down for a closer look. He said afterwards that the thought going through his head at that moment was 'We're in the shit.'

Boss Steve Saint-Amour came over. 'Can you see the problem?' Dave just pointed. Steve adjusted his glasses and bent down. Neither of them said anything; according to Dave, 'We were both of us just speechless. It felt like somebody had just backed over your child.'

At that point Ray Darville came over and followed their gaze. 'Holy cow!' was all he said.

A moment later, I got there. 'What's wrong?' I asked.

Dave looked up and gave it to me straight. 'It's dead, mate. It's a paperweight. The pod's gone – it's been pulped.'

The heart of the vehicle, the electronics capsule, had literally imploded. None of us has ever seen or heard of anything like this before. There are spare parts and back-up components on board for just about every possibility, but not this. The nearest ports are Punta Arenas in Chile and Port Stanley, Falkland Islands, both at least a couple of days away. To fly in replacement parts and steam to either destination would cost us at least a week, which would run us out of time.

At that moment, my prospects of conducting a proper archaeological survey of the *Endurance* also imploded.

John decided to put out a bulletin on the team's internal messaging service. The last line said it all: 'This is a very major blow to the expedition and the search for the *Endurance*.'

The situation was, of course, reported immediately to HQ in London on the other side of the world. Within three hours, they came back to say that they would courier the spare parts to King George Island on a specially chartered plane from the United States. This surprised us all; first, there was the excruciating cost of what was being proposed, and second, none of us had thought of King George Island, which is one of the islands in the South Shetland Chain off the western tip of the Antarctic Peninsula, where the Chileans have an airstrip which connects with Punta Arenas in the Straits of Magellan. The captain was not confident that this was achievable – but it was a plan, we had nothing better, and it might just work.

And so we are heading for King George Island. ETA 0400 hours on 1 February.

It has been a stressful day and spirits are at rock bottom. All the more so because it has now been decided that some of our scientists whose work is largely over will be flown home from King George Island. There are sound reasons for this, but the team has bonded closely and nobody wants to leave while there is still a chance of finding the *Endurance*. None of us yet knows who will be 'on the list'. The mood at dinner was the most subdued I have yet seen.

NOON POSITION: 67° 13.423' S, 045° 52.008' W

31 JANUARY 2019

On our way to King George Island

Today we have been off the very tip of the Antarctic Peninsula on our way to King George Island. This evening we began our crossing of the Bransfield Strait, a body of water about 60 miles wide between the Peninsula and the South Shetland Islands. We are making good time, at 17 knots, through mixed ice, most of which is quite widely spread. Nonetheless we are keeping a sharp lookout for growlers that might slip by the radar unnoticed.

It feels a bit angsty on board, understandably so. We know that 10 of the scientists will be leaving on the first flight out tomorrow but, during the day, we had no idea who they might be as it was a decision made in London. By and large, since the day we left Penguin Bukta it has been mainly big smiles and bags of common purpose. Everybody came together for science and for Shackleton.

During the evening those who will leave tomorrow were informed. There was much dejection and, yes, tears. They all understand why it has to be, but there can be no doubt that this night, 10 of our number were served a tepid dish of damned bad luck.

NOON POSITION: 62° 21.648' S, 048° 33.247' W

1 FEBRUARY 2019

Arrive King George Island. The cruise ship Explorer. *Ten of the scientists disembark. The secret world of ice*

These are *very* dangerous waters – a raceway for bergs being shunted and spun around Antarctica on the circumpolar current. As if to make the point, during the night we passed over the lacerated carcass of an old friend that is rotting out its existence on the bottom of the Bransfield Trough.

Nowadays, tourism in the Falkland Islands (which are slightly east of north of us, across the Drake Passage) is a multimillion-pound industry. But it was all started from scratch by my parents back in the late 1960s, when they launched a small business named Outward Bound Tours. They ran it from my mother's bookshop, which was called – what else? – Boundbooks.

One day I went home and there, seated with my parents, was a man called Lars-Eric Lindblad, a sturdily built, fair-haired Swede who, just a short time before, had led the first tourists to Antarctica. He had recently commissioned a purpose-built Antarctic cruise ship of his own, the *Lindblad Explorer*, which he also intended to use in the Falklands. I don't think there was anything more than a hand-shake between my father and Lars-Eric, but it was the start of a small, informal collaboration in which we arranged tours for his clients when they came to Port Stanley on their way to South Georgia and the Graham Land Peninsula (as it was then called). Several times I met the ship and took the tourists to some of the remote penguin colonies on the West Falklands. During these trips I fell in love with the ship, which, much later, made headlines when she became the first passenger vessel to navigate the Northwest Passage. Lars-Eric died many years ago; the ship changed hands and underwent major refits, and her name was shortened to just

Explorer; but she never stopped taking tourists to Antarctica. That is, not until 2007.

In the early summer of that year the *Explorer* left the Falklands on a special cruise advertised as 'The Spirit of Shackleton', during which it was intended, insofar as possible, 'to follow in the footsteps of the legendary explorer'. During the small hours of 23 November in the Bransfield Strait, when the vessel was just a short distance from the southeast coast of King George Island (at the spot we passed over ourselves last night), an iceberg sliced open her bottom like some old fish on a monger's slab. There was no loss of life but several hours later, engirded by ice and to a sigh of escaping air, she slid under and was seen no more. She was not only the first Antarctic cruise ship, but the first to die in these waters.

By breakfast we were nudging our way up the 12-mile-long Maxwell Bay at the south-eastern end of King George Island, and by 0900 hours we had reached Ardley Cove, where the Chileans and Russians have a base. It is a distinctly bleak-looking place of dark hillsides patched with snow and valleys clogged with glaciers. There are bays full of krill-scoffing humpbacks and minke whales and, along the shoreline, gentoo and chinstrap penguins. With climate change, this island is now more sub-Antarctic than Antarctic.

Before lunch there was a team photo and then, during the afternoon, the 10 scientists left to catch their flight from the Chilean airstrip.

This evening, to help lift our spirits, the bar was permitted to issue everyone with two free cans of beer. I have no fancy words for the mood. With so many of our team now gone (and two more still to depart), it just feels plain old sad.

Where we have been is within the tuck of the Antarctic Peninsula, and thus we were sheltered from the bitter westerlies by its mountainous spine. For much of that time we have been surrounded by a quilt of rather plump, minty-white floes whose movements, though at times quite fast, were always in equilibrium.

However, once the tip of the peninsula was behind us and we entered the rushing Bransfield Strait, all that changed. The ice now comes flying by as if we are on the trackway of the Indianapolis 500. What I've found particularly interesting about the passing floes is that many of them have been flipped by a blast of savage weather which we did not feel when we were within the lee of the peninsula. A striking feature of these overturned slabs, and some bergs, is their now-exposed coffee-coloured underbellies. This isn't of such great interest to others in the team who've seen it before, but to me, in a sea full of frozen mysteries, the rusty staining is beyond fascinating.

It used to be that people thought of the pack as a pointless place, devoid of life, that sometimes sank ships. But this was a mistake, because in the rumpled, golden-brown, crystalline latticework that characterizes the underside of sea ice, there are boom towns of biological activity. In other words, life: glorious, thick and pulsing, and to the animals that feed on it, delicious. This brown sludge, which Shackleton called 'diatomaceous scum', is the beating heart of the Antarctic, a cryological boiler room where the sun's metabolized energy passes from microalgae to krill, who, in turn, stoke the hundred-ton leviathans.

One of the first to decipher this vital, generative natural phenomenon was the great botanist Joseph Dalton Hooker, who was with James Clark Ross in the Antarctic and the Falklands from 1839 to 1843. In his otherwise rather dry *Flora Antarctica* of 1844, he momentarily drops his reserve to enthuse about how 'the waters and the ice of the South Polar Oceans were found to abound with microscopic vegetables [by which he means photosynthesizing microalgae, or diatoms]. Though much too small to be discernible to the naked eye, they occurred in such countless myriads, as to stain the berg and pack ice . . . brown, as if the Polar waters were charged with oxide of iron . . . The universal existence of such an invisible vegetation as that of the Antarctic Ocean, is a truly wonderful fact.'

Shackleton's friend Fridtjof Nansen was also puzzled by the presence of so much life within the frozen wasteland about him. There are, he wrote, 'unicellular pieces of slime that live by the million in pools on very nearly every floe all over this endless sea of ice, which we like to call a place of death. Mother earth has a strange ability to produce life everywhere. Even this ice is fertile to her.'

We now know that what they saw beneath their microscopes were diverse communities of mostly single-celled organisms, living within the matrix of platelets, pores and brine channels that characterize the loose undersides of the floes. These communities consist mainly of viruses, bacteria, algae and protozoa. Some of them feed on each other, but most take their sustenance photosynthetically. All, however, are food for metazoans such as krill, amphipods and copepods, which in turn nourish the birds, fish and mammals. Some day, somebody will write a book entitled *The Secret Life of Ice*, and when they do I shall be first in line to buy a copy.

NOON POSITION: 62° 12.094' S, 058° 56.983' W

2 FEBRUARY 2019

Awaiting spare parts. We go ashore

We've spent all day anchored at King George Island. The mid-afternoon flight with our fresh food, medical supplies and, most important of all, replacement ROV parts, did not arrive because of strong winds and fog. Worse weather is on the way and everybody is becoming concerned because if the flight cannot make it tomorrow, then we will not get away until the 4th.

This place gets me down. Anywhere in a storm but, please God, not here, not King George Island. Conceived by politicians, peopled by scientists, ramshackle and joyless, twinned with nowhere. In the afternoon most of us went ashore in one of the ship's boats for a couple of hours. Everything was a bit stark and craggy, wet and stodgy underfoot and there was not much to see. At the Chilean Presidente Frei base there were some derelict vehicles, the charred remains of a burnt-out building and a lot of trash. At the Russian Bellingshausen Station, literally on the other side of the road, things were a bit more tidy and a tad more interesting thanks to a tiny Russian Orthodox church on the hillside behind. Apart from some chinstraps and gentoos on the beach, that was about the sum of it. Barren and battered, dull and frowning, it's the absolute armpit of Antarctica.

I went to the shop. There was nothing in it that anybody would want. I felt I must buy something, though, as they had opened it specially for me; so I bought two coat-sleeve patches, one for me and one for our ship's doctor, who likes patches. Then I climbed the hill to the church. The priest was there, bearded and in his robes. He gave me a small cardboard reproduction of one of the icons.

In return for their help in carrying some of our people ashore we have been hosting some tourists from a cruise ship, the *Hebridean Sky*, which is anchored close by in the bay. They have been boarding our ship in groups of 10 and our scientists and technicians have been giving them tours of the facilities. I spoke to a couple of them in the lounge and they were most appreciative of being able to come on board. Apparently our coffee was much better than the 'muck' they were given on their ship.

There was some good news, too. I was up on the bridge with Captain Freddie and he showed me the most recent satellite imagery, which revealed some improvement in the conditions over the *Endurance* site. The ice is now slightly more fractured. He attributes this to southerly winds. Whatever the reason, there has definitely been a loosening of the pack, which may let us through.

In the helicopter hangar there is a table-tennis table, but no balls. We have a Chinese student with us called Liangliang Lu and today when he was ashore he walked several miles in freezing cold to the Chinese scientific station in the next bay. His thinking was, if there are Chinese people on this island, there will be table tennis. Sure enough, he came back with four new balls. Suddenly Liangliang is the man of the moment.

NOON POSITION: 62° 12.214' S, 058° 55.661' W

3 FEBRUARY 2019

Still awaiting spares

It is foggy, with low cloud and winds gusting to 25 knots. Nonetheless, twice the plane took off from Punta Arenas – but then twice it had to put back with technical difficulties. We are now becoming seriously concerned. Tomorrow's weather also does not bode well and it feels a bit as if we are swimming in glue.

During the day some of our scientists went ashore for a tour of the scientific facilities at the Chilean base and we hosted visits from the Chilean Antarctic Institute and representatives from Bellingshausen, the Russian base.

NOON POSITION: 62° 12.200' S, 058° 55.703' W

4 FEBRUARY 2019

Still awaiting spares. The celebration of birthdays on the Endurance

We are still at King George Island. All morning there was still low cloud, fog and strong winds. Spirits were about as flat as old champagne.

Normally strong winds disperse fog, but even though the wind speed was 35 knots at lunchtime, the fog remained so dense that we couldn't see the shore or even a vessel anchored 40 metres from us. In the afternoon John and I had a meeting with Captain Knowledge in his cabin to consider the options. He too feels frustrated; as with all his officers and crew, he wants to get back to the business of finding the *Endurance*. We deliberated over the ice conditions, the distances we have to travel and our deadline for arriving back at Penguin Bukta to catch the last flights from Wolf's Fang to Cape Town before the landing strip closes down for winter. Basically, it is essential for us to leave here tomorrow if we are to have any chance of finding the wreck.

We had hoped to conduct a test dive with AUV 7, which will be our principal search vehicle, later in the afternoon but conditions were too difficult. Just as the test dive was cancelled we received a message by radio from Chile to say that, once more, the flight would not be leaving Punta Arenas.

There was an interesting discussion at dinner this evening about what people do for comfort when they're feeling down. I know exactly what I would be doing if I was in London right now; I'd be heading towards my nearest second-hand bookshop for a damn good browse.

However, the day was not entirely without good news: the ice pack over the *Endurance* continues to loosen.

✦

Today is my birthday.

When I started out on my career I was always the youngest on archaeological excavations; I made the tea, cleaned up and did all the 'Hey you' jobs; but now, and for a decade, I have been the oldest, something that really irks me. I know old age must always cede to youth, and so on, but that doesn't make it any easier. So now I avoid birthdays as a penguin does a leopard seal.

This strategy was going well until a well-meaning, eagle-eyed younger member of the team spotted my details on a team list in the purser's office. Birthdays are always celebrated on board with cake and drinks after dinner, so now I feel guilty because all I have done is deprive everybody of cake and an excuse for a bit of a party, which are essential for morale when you're in the icy wastes of the high latitudes.

It was the same with Shackleton's team. Birthdays became important as something they could look forward to, something that would break the monotony, raise spirits and cement camaraderie. Macklin described how Shackleton always marked these occasions with a cake, and once that had been presented he would disappear into the bonded store wherein were kept 'whiskies, rum and dainties' and emerge with a bottle of spirits and sometimes a box of chocolates. As the evening progressed, Hussey would invariably strike up a tune on his banjo.

The King's birthday on 3 June was also an excuse for a celebration. In 1915 Chippy McNish described how they paid tribute to the occasion with 'bread and cheese and butter for lunch and grog in the evening'. Another was the 'birthday' or anniversary of the commissioning of the *Endurance*. By the time they had hauled themselves onto Elephant Island, all the men had left with which to honour their lost ship was a 'tot of methylated spirits . . . with a pinch of ginger'. One of those who had his birthday under the upturned boats on the island was Orde-Lees, who wrote: 'I celebrated my birthday today. We had a fine sledging ration hoosh [a thick stew] . . . for breakfast, for luncheon one biscuit and two sardines each (the last of the sardines) and for supper seal hoosh with four sledging rations added to it, three lumps of sugar and a concert, in which I took part, much to the agony of my comrades.'

It occurs to me that Shackleton's birthday is on 15 February – just 11 days away. When he died he was only 47. By that day, will we have found Shackleton's beloved *Endurance*? Or will we be quietly wending our way back to Penguin Bukta to catch our flights home, having had no sight of the ship? I wonder . . .

NOON POSITION: 62° 12.325' S, 058° 51.013' W

5 FEBRUARY 2019

*We leave without spares. The Polar Medal and why
it was withheld from Chippy McNish*

Instructions from London this morning: 'If the plane does not arrive today, leave immediately.'

It was just what we had been desperate to hear and straight away the mood lifted. These last few days have felt as if we've hit a pocket of dead air and every last drop has drained from our sails.

During the morning the wind lessened and the sea subsided to the extent that we were able to dive AUV 7. The objective was to test its photographic capability, which has given cause for concern, and also to experiment with a more tightly controlled ascent, since the ice openings within which we will be working are likely to be squeezed. The tests went extremely well and at 1115 hours the AUV was back on board and in its cradle.

At 1600 the weather improved dramatically and soon afterwards we heard from Chile to inform us that the flight from Punta Arenas with the spares was on its way. At 1915, however, as we were all eating, our personal radios crackled to life with a message from the bridge informing us that the plane had had to turn back after completing well over half its journey.

What does this mean in terms of achieving our objectives? It means we can still conduct our search for the *Endurance* – but if we find her, my dreams of conducting an archaeological survey of the site are as dead as our ROV.

It also means that if AUV 7 gets into trouble, we will not have an ROV to conduct a rescue mission.

At 2050 we left this rather unlovely little island. None of us were sorry to see it disappear in our wake as we set off across the Bransfield Strait and back into the icy embrace of the Weddell Sea.

If all goes well we will be back at the ice edge on 8 February, over site on the 11th, conclude the search on the 14th, and return to Penguin Bukta on the 20th to disembark.

The feeling of elation throughout the ship was almost palpable. This really is the greatest wreck hunt there has ever been, and once more, after five dead days, the game was afoot.

Everybody was in the mood for a celebration and fortunately we had the perfect reason for one. Yesterday evening we learned that the Queen had been 'graciously pleased' to confer upon our expedition leader, John Shears, the Polar Medal. This is something well deserved and we are all, of course, happy to bathe in his reflected glory. John began his career at the British Antarctic Survey in 1990 as its first environmental officer, and by the time he left in 2015 he was head of operations and logistics.

This evening we all gathered in the main lounge, where we were each allowed two cans of beer or two glasses of wine. I said a few words of appreciation before proposing the toast and then our chief scientist, Sarah Fawcett, presented John with a large mock Polar Medal (so he could practice for the real conferment at Buckingham Palace) and Harry Luyt, dressed as Shackleton, presented him with a certificate signed by all of us.

John and I had an interesting conversation later on about why four of Shackleton's men were not awarded the Polar Medal. The omission of the carpenter, Harry 'Chippy' McNish, was a particularly curious business.

In its current guise the Polar Medal goes back to 1904, when it was bestowed upon the members of Scott's *Discovery* expedition. Until the 1960s it was awarded to just about anybody who had been on a government-endorsed polar expedition; since then, however, the emphasis has been on personal achievement. During Shackleton's time the conferment of the medal fell within the gift of the expedition leader, so there can be no doubt that Chippy's omission was a deliberate snub. To some extent it is understandable: the old Glaswegian was the only one who openly rebelled against Shackleton's authority, buckling under Worsley's instructions and refusing to assist any further in dragging the boats across the ice. His assertion was that once the *Endurance* had gone down, the ship's articles, which they had all signed, no longer applied and therefore they were no longer obliged to follow orders. Shackleton, of course, was furious. He read them all the ship's articles, which included an insert he had made stipulating that the crew agreed 'to perform any duty on board, in the boats, or on the shore as directed by the master and owner'.

Chippy was not in good health. He suffered from piles, and there may have been other ailments, and his arguments concerning the futility of the exercise and the possibility of straining the boats were not without merit. At the time he held a grievance against Shackleton for having had his beloved cat, Mrs Chippy, shot, something which we know was a source of anguish to him; but ultimately McNish was an eminently practical man, and I'm not convinced that the cat figured largely in his act of defiance. Nonetheless, whatever the underlying reasons, open insubordination on that level was not wise and Shackleton had no choice but to stamp it out in the forceful manner he did.

The incident was entered in the log at the time but, interestingly, Shackleton had it struck off during the voyage of the *James Caird* to South Georgia when, according to Shackleton, Chippy showed both 'grit and spirit'. This is when, perhaps, Shackleton should have put the rift behind him, but clearly Chippy's moment of rebellion had been neither forgotten nor forgiven.

The other three who did not receive the medal were John Vincent, William Stephenson and Ernest Holness, all fo'c'sle hands. One of the ship's doctors, Alexander Macklin, wrote of them: 'They were perhaps not very endearing characters but they never let the expedition down.'

Vincent was the strongest man on board. He had been an amateur boxer and was at times rather threatening to his shipmates unless he was given his way. He had been signed on as bosun, but after complaints of bullying he was summoned to Shackleton's cabin and demoted to able seaman. Later, he lost part of his upper lip when it froze to a metal cup, and on one occasion when he fell into the water from the *James Caird* he refused a change of dry clothes because, according to one diarist, 'it was freely stated that he had a good deal of other people's property concealed about his person'. During the crossing to South Georgia in the *James Caird*, his health deteriorated to the point where he became useless. It is unclear why Shackleton selected him for that final haul. Presumably it had to do with his physical strength, and certainly Shackleton had need of his seafaring skills; but it may also have been that he was seen as a potential troublemaker who could not safely be left on Elephant Island, where everybody had to pull together if they were to survive.

Why Stephenson and Holness were denied medals is more puzzling. Neither of them appears to have done anything that warrants the kind of shaming that such a refusal represented. As the *Endurance*'s firemen – the ones who shovelled coal into the boilers – they were at the very bottom of the ship's social ladder, so there may have been an element of snobbery to their exclusion. Little is known of Stephenson, but of 'Ernie' Holness we know more: he was the unfortunate one who, still in his sleeping bag, famously fell into the sea when the floe opened up beneath his tent. He was hauled to safety by Shackleton himself before the ice could come together and crush him to pulp. During the dunking he lost his tobacco, which was hard on him because, like many on the *Endurance*, he was addicted to smoking.

After the war, Holness returned to fishing in the North Atlantic. In 1924, aged 31, he lost his life when he was swept overboard from a trawler off the Faroe Islands.

NOON POSITION: 62° 12.305' S, 058° 51.063' W

6 FEBRUARY 2019

Crossing the Bransfield Strait and back among the bergs.
Life under the Weddell Sea and the wreck of the
Endurance *beneath the pack. A gift*

I was watching everybody having a good time in the lounge last night. The Kafkaesque nightmare of recent days is now behind us. We all feel like Antarctic veterans and can't wait until we are again plying our way through the white hell of the Weddell Sea.

We are striding along at 16 knots near the tip of the Antarctic Peninsula, where great waters conjoin and the continent dwindles to nothing. In these parts no ship ever follows a straight line, it's all warps and wefts, and right now we are performing a grand parabola as we move around the pack from west to east in order to get south.

In all directions our view is dominated by bergs that range in shape and size from soaring, pointed multilaterals and long, vertiginous flat-tops to smaller thimbles, pinnacles and broken teeth; and finally there are the plumper and more bosomy bergs whose edges have been softened by snow. Rejects all, they have been slung from the Weddell Basin by the gyre and are now working their way northwards into more temperate regions where they will expire – water to water. Overhead this morning there were huge, fleecy clouds that afterwards turned leaden and then smothered us in flakes, so that by the evening our decks were bridal white.

The best news is that the latest reconnaissance from space shows the pack over the *Endurance* continuing to open. Satellite imagery was extremely useful to us in our navigation of the gap between the Larsen C ice shelf and the giant A-68 breakaway iceberg, and it will again be helpful when we force our way through the pack into the *Endurance* search area. The person in charge of satellite analysis is our glacial geophysicist Frazer Christie, an affable, six foot six, brainy-looking, jet-black-haired Aberdonian post-doc of 26 years from the SPRI. During the evening he explained that we are downloading data from 12 different satellites on a nearly daily basis and that our sources represent a range of national and international space agencies. When we combine it all, we have a truly unprecedented understanding of the sea ice conditions ahead of us. No other polar project, Frazer says, has ever made such extensive use of satellite intelligence.

In other oceans I've gazed upon wooden wrecks that are deeper than the *Endurance*, but these were on seabeds with which, to some extent, I was familiar. By contrast, what goes on beneath the permanent hard roof of the Weddell Sea is literally uncharted territory.

On the wall in my day room is a large hydrographic map of the Weddell Sea: the well-known Admiralty Chart Number 4024. The greatest danger to ships at sea, apart from fire, is land, and usually the nearest land is that which is under your keel. Knowing the depth of water beneath you is, therefore, fundamental to safe navigation. On a chart the depth is usually expressed in a series of contour lines, or isobaths, which give you not just the depth but the topography, or relief, of the ocean floor. And this is where my chart of the Weddell Sea becomes interesting. The contour lines are there and follow the normal conventions, but when you come to the sector which is permanently covered by pack – that is to say, the part of the Weddell Sea where we are now – there is nothing. The contour lines stop dead in their tracks. There aren't even any conjectural dots. It's just blank and void. Right now, we are, literally, off the map. Nobody has been to the bottom of the Weddell Sea.

So what is it like in the 3,000 metres below us? An extreme world, certainly, but also a dull one. These, after all, are the mudlands of the Weddell Sea, a vast Serengeti of terrigenous silt. A world without light and drained of colour, just earthen hues and dark dilutions that blend one into the other. Apart from drop-stones from icebergs and the occasional geological outcrop, features will be few.

That is not to say there is no life. In fact it is pulsing with bioactivity; spectral planktonic drifters, dark gelatinous organisms and cryptic little dots and dashes that reside within the seabed. And, of course, there will be microbial communities that barely qualify as life. If there are free-swimming fish colonizing the wreck, they will be nothing like the sassy show ponies one finds on warm-water coral reefs; more likely they will be related to, or reminiscent of, the crocodile ice fish, ghostly white from the absence of haemo-globin in their blood. And finally there will be what the Italians might call the *signori di notte*, the lords of the night: the giant squid. To see one of those would make my life complete.

During the period that the *Endurance* was icebound, Robert (Bobbie) Clark, the team biologist, was forever dredging the seabed. Frequently he was rewarded with what they called 'a good haul'. We do not know much about the living organisms he recovered, but Clark kept his specimens in sealed honey jars and these are still on the wreck. We will not be touching the *Endurance* but if those jars are just lying there exposed on the seabed, a case could be made at some point in the future to recover at least some of them for study and to compare their contents with their modern-day counterparts.

As for the wreck site itself, that could be like an oasis in the abyssal desert. Unlike the Sutton Hoo ship, which survives only as a ghostly archaeological imprint in the dirt, ours will be very real and wholly evident. It is the kind of site that will offer all sorts of nutri-tional opportunities for deep-sea fauna, and it will have any number of little crannies all perfect for the accommodation of lesser and more vulnerable life forms. From an archaeo-biogeochemical perspec-tive, the *Endurance* is important because she represents the only human construct to exist beneath permanent pack, and the only wreck at the heart of a sea that is otherwise without wrecks; but more than that, she embodies an untouched artificial reef of precisely known date which has evolved into a unique ecosystem. Within that ecosystem there is a continuous exchange of chemical elements and simple substances between the colonizing organisms, the ship itself, its natural setting, and the perpetual snowfall of death, dung and particle matter from the three kilometres of water above.

Sitting at my desk in the day cabin earlier, I heard a polite tap and then – a bit like a seal popping its head up through a hole in the ice – a face appeared from behind the half-opened door. It was Kurt Spence, one of the students from the University of Cape Town. 'I have something for you,' he said; and then he produced a small wire-frame model of the *Endurance*.

Kurt is a metalwork artist. When his duties in the lab are done for the day he sits with a soldering iron, making things. When I watch him at work he reminds me of 'Jimmy' James, Shackleton's physicist, who was later also at the University of Cape Town. Like Kurt, he was forever fabricating things out of tin and wire. His cabinmate on the *Endurance* described him as 'a perfect pest for untidiness' who had 'some facility with a soldering iron' and was 'always constructing things of his own invention. He made a candle extinguisher and now he has made a thing that looks like a toy engine [but] is really an electroscope.'

Kurt's model is stunning. It is only about seven inches long, yet it displays all the ship's main features in their right proportions: the funnel dominates, the vessel's four little boats are there on their davits, and it even has a petite propeller that turns. To cap it all off, he has set it upon a piece of wood he found on the beach at King George Island. It must have taken him several evenings to make. I am very touched.

NOON POSITION: 61° 58.172' S, 050° 48.745' W

7 FEBRUARY 2019

'There was a ship,' quoth he. The Endurance *search box.*
In heavy snow we circumnavigate the pack

As a boy, I read nothing but stories of the sea. The Hornblower books were my particular favourites. My first serious job after leaving school was as a greaser in the engine room of an old tramp steamer in the South Atlantic. She had a company of characters that would have left any Hollywood casting director frothing with envy. There was a genial if distant skipper, a brutal mate, a drunken first engineer who never left his cabin, a cabin 'boy' who was 70, a highly strung steward with, let's say, some unusual proclivities, and so on. Since then, as a maritime archaeologist, I have served on a long line of ships – each different but each, always, with a full hold of ship-dwelling zonks and nautical oddballs the like of which you do not find in any other walk of life. To mint an adage, normal people do not go to sea.

I have gone through typhoons in the South China Sea and hurricanes in the Caribbean, and I have been holed up on a reef off Ecuador while surrounded by sharks, all licking their lips. In each of these cases, there was a good chance that we might die. Ships can give you moments of subliminal happiness but they can also stretch you in ways you never thought possible – and then there are those extreme situations when they just rip your very guts from you. They bring out the best, the worst and the most downright freaky in people.

The 28 men that went south on the *Endurance* were a ragtag ensemble of adventurers, scientists and sailors who, in the words of Jimmy James, had been brought together 'with a single objective, to cross Antarctica'. That dream ended before it began; instead of crossing the continent they became men locked in a box, who were then cast adrift on the ice, after which they stared down death from open boats on vast seas before finally being marooned on one of the most hostile islands on the planet. Whatever their initial goal may have been, in the end it came down to just one thing: getting out of there alive. Hurley summarized it well when he wrote: 'Here we were, a small coterie of scientists and seamen, poles apart in learning, culture, beliefs, and creeds, banded together with the one common impulse – to survive.'

The story of what happened to those men after they left England has been shaped almost entirely by *South*, which was presented as a factual record of their epic journey. The book appeared in 1919, three years after they got back from Antarctica, and (the ultimate accolade in publishing) has never been out of print since. Back in the '60s, when I first read *South*, I was happy to go along with its every word, but now I am not so naive – now I want to get closer to the truth of what happened. *South* was all about myth creation. Shackleton was hungry for fame and fortune and to be seen and adored as a hero. This he had achieved with the publication of *The Heart of Antarctica*, about the *Nimrod* expedition, in 1909; but that book was later knocked into the shadows by the 1913 publication of Scott's harrowing diaries, which established him as the pre-eminent British explorer of the era. Scott might have been dead, but the bitter rivalry between the two continued. *South* was Shackleton's last chance to carve his name in legend and, if not eclipse Scott, at least ensure his place beside him.

For books of this type there was a template which, fortunately for Shackleton, perfectly matched the vision of himself that he wanted to project: the noble hero, the fearless leader of a loyal, selfless, body of men who, inspired by his example, worked seamlessly for the good of all. In other words, the perfect expression of the Imperial ideal. Certainly there was much to admire in those 28 who went into the Antarctic on the *Endurance* – they were a remarkable group, capable of great fortitude and generosity of spirit, and they were in large part loyal and brave. But, as the diaries reveal, they could also be snarky, nitpicking, jealous, self-serving and vindictive. They were not, every one of them, chin-up and valiant-all, and they were definitely no band of brothers. In short, they were – as my mother used to say when I was venting my frustration over some bonehead who was complicating my life – 'people, dear. Just people.'

Until today, I thought I was the only one on the ship who was troubled by doubts. The closer we get to the ice edge, the worse they become.

Talking with colleagues at lunchtime, though, I learned that I am not the only one feeling this way. Their main preoccupation is whether we can actually break through to the *Endurance* site. On this I have not the least fleck of concern; having studied the satellite imagery with Freddie and having seen what this ship can do, I know we will get there. But I have two overriding anxieties of my own, one of which in particular is now keeping me awake at night.

The first is equipment performance. The latest generation of AUVs are truly incredible, but I have been around them for quite a while and have seen them self-abort their missions many times. What happens, I ask myself, if they do that here, beneath the carapace of the pack? That would present a situation far more challenging than when we lost AUV 9 under the floe. Furthermore, we no longer have a ROV to perform a recovery if needed. Any way you look at it, this is going to be the most punishing mission ever undertaken by a modern, non-military AUV. And yet, I do not spend much time worrying about this because there is nothing we can do about it. AUV 7 has gone through its wet tests and its payload systems could not have been more rigorously checked. In other words, this one is now in the lap of the gods.

But then there is my second concern: what if the wreck is not in the search box? And the responsibility for *that* would lie entirely in my lap.

The search box is of my designing, and within the next 36 hours I must sign off on it. In these matters I consult closely with my colleague Toby Benham at the SPRI – even today the emails have been flying back and forth between us, shaving off a mile here, adding one there – but in the end, it is all based on my research and the final call is mine. Everything derives from that one key bit of information: Worsley's position taken the day after the *Endurance* sank. On the surface, all this looks fairly straightforward – but it is not.

Several years ago, there was another attempt to find the *Endurance*. It never reached mobilization but later I was shown their intended search box; they had simply taken Worsley's position, made an allowance for drift, and then drawn a square box around it. This told me straight away that they had not done their homework. During the preliminary stages of our own project, I used to say, 'Show me a map and I will put my finger on the *Endurance*' – implying that she was not lost and that I knew exactly where she was. But the truth is more complicated than I first realized, and the more closely I scrutinized the problem, the more rubbery everything became.

Normally in deep-water wreck-hunting, you begin by defining your survey box. You surmise the most likely location based on your research and then you draw your search area around it. This 'box' takes into account your margin of error, and very often you balance it more to one side than the other to reflect the greater likelihood of where your target might be if the evidence has played you false. From this you calculate the area, or acreage, of your search and, once you know that, you devise a schedule and address all of the logistical and operational considerations that comprise your mission plan.

With our hunt for the *Endurance*, however, everything is a bit back to front.

The days we lost at King George Island were a disaster. Thanks to head office in London and the generosity of the Flotilla Foundation we were able to extend the ship's charter, but we have now reached the very back end of the season and are on borrowed time. Winter is upon us; the airstrip at Wolf's Fang on the eastern shoulder of the Weddell Sea is about to be closed for the season, and the *Agulhas* has other charters and obligations that she must fulfil. By the time we reach the site we will have only 50 hours of search time. This means one full AUV dive of about 42 hours followed by a second dive of about eight hours. In other words, the square acreage or amount of terrain I would like to cover has been significantly compressed, so I must now draw up a new search box that reflects our reduced 'bottom time'.

How I manipulate the fixed acreage I have got, to arrive at the dimensions, orientation and placement of the box, comes down to a number of factors that have to be expressed in lines of latitude and longitude. These are usually fairly straightforward determinations – but not this time, as they pivot upon drift, chronometer error, sextant accuracy and other considerations, some of which are rather abstract and to a certain extent imponderable. Normally this would not worry me too much because you simply ensure your search box has the necessary margins to absorb all the uncertainties; but in this case that's not easy, because the search area imposed on us by circumstance is so small. The amount of seabed we can cover in 50 hours is really tiny. Every mile matters and could be the difference between finding and not finding Shackleton's ship.

In any search for the *Endurance* the starting point has to be Worsley's coordinates for the sinking: latitude 68° 39' 30" S, longitude 52° 26' 30" W.

However, when I began my research on the loss of the ship I soon realized that things were not as they seemed. Everybody assumes those coordinates are a precise celestial position; they are not, they are an estimate. And that gets us to the heart of the problem. Because of overcast weather conditions, Worsley was unable to obtain a sextant sight during the three days before the ship sank, and as we know, his first observation after she sank was on the following day. The wreck sank on 21 November 1915; Worsley's last position was on 18 November and his next was on the 22nd. His problem (now mine) concerned the speed and direction of drift during that period; also bearing in mind that he was

not by the wreck when it sank but at Ocean Camp, about a mile
and a half from the wreck in a roughly northwest direction. So to
obtain his famous coordinates for the sinking, he took his observed
position of 22 November and applied what is called an offset.

My search box consists of a north and south line of latitude,
and an east and west line of longitude. In very simple terms, calculating latitude (the notional lines that run east to west around the
globe) requires the use of a sextant to measure the angle between
the horizon and the sun at noon, when it is at its highest point.
Calculating longitude (the notional lines, or meridians, that run
from pole to pole) is more challenging because it depends upon a
calculation of time. That is the difference between local mean time
(LMT) and Greenwich mean time (GMT, the zero meridian), which
in Shackleton's day had to be carried with you on chronometers.

The northern line of latitude of my search box was derived
from Worsley's latitude for the 22 November. The weather was
clear that day, and for somebody of his skill, shooting the noonday
sun would have been a simple matter. Any way I looked at it, I
could not see how the wreck might be north of my search box.

The southern, or higher line of latitude, was a bit more complicated to work out, as it required some informed guesswork on the
speed and direction of drift. From 6 to 8 November the men had
experienced a 'howling' blizzard after which the winds (the main
influencing factor on drift) were predominately northerlies. Because
they were depending on the pack to carry them north towards the
mouth of the Weddell Sea, any wind coming at them from that
direction was unwelcome. Although, as plot charts for the *Endurance*
demonstrate, this was a period of extreme confusion in the drift
of the ship, and they were not pushed as far south as they feared.
On 18 November, when Worsley took his last observation before
the sinking, one of the diarists noted that 'we have not been driven
back more than a total of 3 miles since the blizzard'. And on
22 November, the day after the sinking, Worsley was pleased to
note: 'Although we have had the bad luck to have had some 10 days
of NE breezes tho' mostly light, we have the satisfaction to find

that the Pack shows great resistance to the S.W., at all events we do not move more than 5 or 6 miles to the south in that time.'

From the position on 22 November forward, a steady northerly drift was re-established. We also know that the winds on the day of the sinking, 21 November, were variable southerlies. Therefore the greater likelihood is that the ship was also trending north on that day. It did not matter how I looked at it, I could not see how the *Endurance* could be more than nine nautical miles south of Worsley's observed latitude for 22 November, given that the drift speed rarely went above 0.4 knots and that her average daily drift was four nautical miles. In summary, when I drew up the search box I was confident of its lines of latitude.

In a lifetime spent investigating wrecks, I've always found longitude to be the problem. A miscalculation of longitude is believed to have been the cause of one of the greatest disasters in British maritime history when, in 1707, Vice Admiral Sir Cloudesley Shovell's fleet drove upon the rocks of the Scilly Islands resulting in the loss of four ships and 2,000 lives, including that of the Admiral. It was a wreck area I first viewed when I was a young diver and journeyman archaeologist, and it was while inspecting the debris from this tragedy that I first became interested in the problems of navigation at sea. What happened on that site led to the establishment of the Board of Longitude, the purpose of which was to encourage the most inventive thinkers of the day to apply themselves to solving the problem of longitude – something which was not achieved until John Harrison's invention of the marine timekeeper some 20 years later.

With longitude, everything depends on the accuracy of your chronometers, and a century ago they rarely kept precise time. In any calculation of longitude there were, therefore, two major areas of vulnerability: the first was local time and the second was Greenwich time. Because chronometers were so important, the *Endurance* departed England with no fewer than 24 of them, all of which were rated in London before they left and again at Buenos Aires on 24 October 1914. After leaving South Georgia in December

1914, they never saw another landmark of known longitude until they espied the mountains of Joinville Island in March 1916.

Chronometers were notoriously sensitive and had to be looked after with great care. Physicist Jimmy James wrote: 'We were actually out of sight of any land whose position was accurately known, and could thus have served as a check on our observations for about 15 months. During that time the chronometers had been subject to variation of temperature, and part of the time had been on a ship which was forcing its way through the ice and were thus liable to receive sudden bumps.'

In the end, out of the 24 chronometers that left England there were only three that Worsley believed still to be reasonably reliable. Of these, the only one he considered to be in 'good going order' was the one he wore around his neck beneath his sweater, for the sake of protection and to ensure an even temperature. This was the famous Smith's chronometer that Shackleton later gifted to Worsley and which is now on display at the Scott Polar Research Institute.

During the darkness of their first Weddell Sea winter, when the *Endurance* was locked in the ice, Worsley and James began observing occultations to check their chronometers. An occultation is when one celestial body eclipses another, the precise times of which were known to them from a nautical almanac. That first night in June 1915 they recorded the occultation of Sigma Scorpii followed by three other stars. From these observations it was evident that the accumulated error on their chronometers was significant. James wrote: 'The chronometer errors calculated from all these agreed within a few seconds, and it was nearly 6.5 minutes! Over a degree and a half in longitude, perhaps 20 miles at that latitude.'

Occultations continued through to September (when the long nights of winter ended) and it is clear from the diaries that James and Worsley were quite pleased with their results. Geologist James Wordie, who later made a study of the drift of the *Endurance*, wrote that 'when more occultations came to be observed in July, it was possible to fix new rates for the chronometers, and the August and September observations accordingly never involved more than slight changes.'

Six months later, when Joinville Island was sighted and bearings could be taken, they found that their calculated position was only 10 degrees too far to the east, or about 3 miles – which, all things considered, was not bad.

In broad terms, when defining the east and west extremities of the search box, I had three choices. Either the chronometers were slow and I had to weight my search to the west, or they were fast and I had to weight the search in the opposite direction; or thirdly, their occultation work was good and there had been relatively little error creep between the last occultation in September and the sinking in November, in which case Worsley's longitudinal estimate for the sinking would be fairly accurate. The current preponderance of scholarly opinion is that their favoured chronometer was slow; but there is also evidence to suggest it was fast. Which raises yet another problem, which is that we do not always know which chronometer they were using. In the end, the search box I drew up was roughly oblong in shape, aligned east to west and weighted to the west.

And now comes a twist. I claimed earlier that I had faith in my lines of latitude. That was true, but since entering the pack that confidence has taken a shaking.

By kind courtesy of a colleague, Simon Gilbert, I have been able to bring with me a sextant of the same year, manufacture and type as that owned by Hubert Hudson and used by Worsley on his crossing to South Georgia. This is the one the mate saw me with on the deck earlier in our voyage. I brought it because I wanted a better understanding of the navigational challenges faced by Worsley. Before leaving England, the sextant was serviced and its defects made good by a specialist and then myself and another colleague, Mike Critchley, took it to a beach to test it by shooting a few sights. We experienced no trouble in obtaining our altitudes and I therefore anticipated that it would be the same when I got into the pack. But it wasn't.

I have now taken the sextant out on the floes twice. Both times, I was unable to determine an even skyline to which I could bring down the sun to the point where its bottom edge (or so-called 'lower limb') just touched the horizon. My view was obstructed by great pulpits of ice, pressure ridges and, of course, the ubiquitous bergs. I had been aware of this issue from the passing comments of some of the diarists, in particular the physicist James who, in his own words, 'had a lot to do with the position finding of the expedition' and who was actually more informative and reliable than Worsley when it came to discussing methodology. James, for instance, observed that 'In thick pack the difficulties of position finding with a sextant are sometimes considerable, for the instrument depends on a visible horizon. The horizon in hummocky pack may be rather irregular and uncertain.'

As Worsley himself bluntly put it: 'If you cannot see the horizon it is impossible to measure the altitude of the sun to establish your position.' I was not, however, concerned that this would impact the reliability of the positions Worsley took at Ocean Camp (from where they witnessed the sinking of the *Endurance*), because he was taking his observations from a 15-foot-high, wood-built 'lookout' station which would, on a clear day, have given him an unrestricted view of the horizon. But since entering the pack myself I've become aware of a more serious complication: refraction.

I knew of the problems caused by refraction from the writings of the diarists and Worsley himself, but I myself had never seen it before – and thus I had no idea of how badly it could obfuscate and distort the line of demarcation between earth and atmosphere. Since entering the pack the night before last, I have been out on the deck with the sextant many times, and not once have I been able to divine the true skyline. I am not talking here about a misty horizon – which is extremely common at sea – but rather a sort of visual falsification, in which bergs appear stretched and elevated and where everything is lost in shimmering, blurry, non-linear deformations caused by the bending of light.

This type of problem was commented upon several times by the diarists, especially when the sun was not very high above the horizon. James, for instance, wrote: 'Its [i.e. the sun's] altitude may still be rather badly affected by refraction errors . . . all one can do is to rely on a single observation of the sun at noon, and correct for refraction as well as one can. But the possibility of error remains quite high.'

What all this comes down to is that my confidence in the search box has been shattered by the refraction I have seen. The cornerstone of my thought all along has been the position Worsley took on 22 November; but if he was trying to take an angle through such a swimmingly extravagant horizon, I cannot see how I can any longer consider it a reliable navigational observation. In a search box as small as ours, a tiny error can make the difference between success and failure.

And so, all of a sudden, doubts – great, fat, suppurating doubts – are slamming about within my head like breasting sea elephants upon the beaches of South Georgia. Shackleton spoke of the 'quiet hours of the night, when you think over the little snakes of doubt twisting in your heart'. 'I have known them,' he wrote; and now, so have I.

If the *Endurance* is not in the box, she should be very close to it. On this expedition, though, 'very close' simply isn't going to cut it. Usually in operations of this kind, if the target is not there, we just extend the search box, but this time we've only got 50 hours to work with. If we don't find the wreck in that time, there will be a blizzard of rebuke.

I have been in a situation like this before. Four years ago, I spent five months searching for Admiral Graf von Spee's lost First World War fleet. I didn't find it, and in that moment everybody had a view about where I should have looked. None of them had done the research, yet suddenly they were all experts. That experience was a bit wounding and I don't especially want to go through it again.

There is, however, one major difference between then and now. During the search for Spee's fleet, relatively few people knew what we were doing; this time, the eyes of the world are upon us. I can just imagine the scolding I will get if we fail.

Last night we were engulfed by snow. It began about midnight with a few aimless flakes and then, over the next hour on a growing wind, evolved into a swirling frenzy, after which the heavens darkened and it came down in a great, wet, oppressive dump that was so dense we couldn't see halfway from bridge to bow. Within minutes, our decks were blanketed calf-deep and all the machinery was lost in drifts a metre high. By morning the weather had cleared and, although there were still bergs everywhere, visibility was good and we were motoring along at a decent pace in a southerly direction. We have circumnavigated well over half the rim of this great scrotum of ice that abuts the lower Antarctic Peninsula, and now (to mix metaphors) we are going to rip our way up through its soft underbelly.

I spent half an hour with the ice skipper, Captain Freddie, studying NASA's MODIS (moderate resolution imaging spectroradiometer) images from its Terra satellite. The 'patch' we've been observing, with its maze of leads, is still there and, if anything, slightly more open – all of which is excellent news as it will help in our approach to the site. But we can also see areas where there is ten-tenths coverage and where we will have no choice other than to use this ship like a ball-headed club of war to bludgeon our way through.

In a little over 24 hours we shall reach the ice edge and then we will be into the pack. As the mate said, with his usual crisp choice of words: 'We are heading into Hell.' The only thing of which we can be certain is that our way will not be paved with good intentions, only ship-hungry, hull-lacerating ice. There is danger and, one way or another, something cataclysmic, defining and massively decisive is going to happen; but we have no choice. We have to go in. And when we come out I suspect that, at some level, we will be different people.

NOON POSITION: 65° 40.679' S, 044° 59.989' W

8 FEBRUARY 2019

Into the pack. Bickering within Shackleton's team.
Swarmed by seals

Society on the *Endurance* was every bit as stratified as back in Britain. The crew (or the 'lower social orders', as one of the diarists quaintly put it) knew their place – they did not eat with the others and they were much scorned by some of the educated elite on board. They were excluded from the famous midwinter festivities in which men dressed up, sang songs, performed music-hall routines and toasted the King; they were given a tot of rum and sweets, but not invited to this very special occasion. We know they felt aggrieved by their exclusion and you cannot help but wonder what they said among themselves. Since none of them kept diaries, we will never know.

Despite the impression of cohesion and singular purpose given in Shackleton's account, as one might expect with so many strong personalities, there were simmering resentments, open animosity and even a fistfight. The diarists talked of 'conflicting interests', 'estrangements', 'rows', 'bickering', 'differences', 'friction', 'contretemps', 'arguing', 'squabbles', 'altercations', 'snappiness', 'clashes', 'heated arguments' and 'bursts of temper'. 'We are all so ludicrously touchy,' wrote one man. There was bullying (the ship's bosun was demoted because he called the fo'c'sle hands 'evil names and struck them') and some who were unpopular were even victimized. Green was called 'the stupid cook', Orde-Lees was described as 'a perfect pig in every way' and Chippy McNish was 'a horrid old man'. McNish, for his part, held many of his shipmates in complete scorn, particularly the 'ones who have never don a days work in this world [sic]' – these he called 'pharasites' (i.e. parasites). Hurley was also scathing of those 'who shirk duties', implying that there were quite a few of them.

During their time on the ship there had been flashes of temper, but this was far from the internecine squabbling that came to characterize relations between certain individuals and groups once they were on the ice – more especially when they were on Elephant Island, where their existence became, as one diarist put it, 'a living death'. Unsurprisingly, discord increased with the passage of time, but there was a particularly sharp rise following the slaughter of the dogs. Without the animals to feed and exercise there was nothing left to do except read, play cards and suffer the cold: fertile ground for the growth and multiplication of disagreements.

For instance, in March 1916 Macklin wrote: 'This morning there was quite a lot of unpleasantness on rising; Clark and I came to high words, and Worsley and Lees had a wordy get to also. Greenstreet upset his milk and, so exasperated was he, that he turned on Clark . . . blaming him as the cause of it.' People's personal habits began to annoy, leading eventually to anger or disgust; as one man put it, 'certain people do have undesirable idiosyncrasies that do get one's nerves down'. Again, it was Macklin who painted a more fulsome picture: 'Clark has an almost intolerable sniff – he sniffs the whole day long and almost drives one mad when one has to remain inside with him. Lees and Worsley do nothing but argue and chatter about trivial matters and the rest of us can do nothing to escape from it. Lees at night snores abominably . . . altogether one has to endure a good deal . . . with Clark sniffing into my ear, my only relief is to take up my diary and write.'

James also talked of the 'occasional quarrels' and the antagonisms that arose out of very little: 'Do what one will, one comes at times to hate the very sight of some particular unfortunate from whom one cannot escape. The way he looks, or eats, or walks, or sniffs, becomes abhorrent, and one has the greatest difficulty in refraining from telling him so, and is not always successful. Usually one finds someone else who has similar views, to whom one can unburden oneself. Then the antipathy passes away completely, and a new one develops, and new alliances arise . . .'

The problem with the formation of small cells of like-minded individuals within a team is that they often look for targets, which are never hard to find. This can then lead to bullying. Orde-Lees was victimized in a particularly unpleasant manner. He knew it when he wrote: '. . . one cannot deny that there is a very undesirable tendency to cliques. I believe this always happens on polar expeditions. I often think that there is a clique operating against me. I wonder if this is really so, or whether it is only an hallucination of one's distorted brain.'

Orde-Lees was certainly unpopular and there can be no doubt that he was treated with spite. Macklin, for example, describes how, when they were on Elephant Island where tempers were short, people became mean and everybody was ravening, so that Orde-Lees was not given his 'fair share of food' for being late. On another occasion while upon that abhorrent rock, they used his bag to protect the stone floor from becoming dirty while skinning penguins; thus when he came in he found it completely befouled. 'I was not prepared for the frightful mess in which I found my bag . . . no one knew anything about it.' A story was told (which appears to have had no basis beyond a few loose words spoken years later by the cook, Charlie Green) that when the food ran out on Elephant Island, Orde-Lees was to be first for the pot.

Asleep in her bunk shortly before midnight on 14 April 1912, the Countess of Rothes was suddenly awakened by 'a slight grating sound'. She drifted back off to sleep, not realizing that what she had heard was the noise a ship's hull makes when it is being sliced open by ice. That ship was the *Titanic*, and – of course – she was not asleep for much longer.

I thought of the Countess as I too awoke during the night to the sound of scraping ice. By contrast, however, I knew immediately what it meant: we were now into the worst patch of the worst portion of the worst sea in the world. We were into the pack, and somewhere below us was the last resting place of the *Endurance*.

I got dressed and went up on the bridge. There were only four people there: Captain Freddie, the second officer, a deckhand, and Liangliang on ice monitoring duties. This being the austral summer you could see almost as well as during the day, even though the sun was low on the horizon. Apart from a narrow lane astern, we were right at the centre of an icescape that stretched as far as the eye could see in every direction. Occasional clumps and the odd small berg cast long shadows; otherwise it consisted almost entirely of lean, flattish first-year ice that crumpled easily beneath our bows.

From the pack edge to the search box it was 115 nautical miles. We were currently doing 5.5 knots, moving roughly with the drift on a heading of 336 degrees, or almost nor'-nor'-west – although precise headings often do not mean very much when you are within a frozen crust like this, hopscotching about between potholes and leads. Very soon, however, we shall be out of the current field and into the multiyear stuff, which will curtail our progress because of its greater thickness and durability. Luckily there are many openings ahead (which is why we chose this route), so as long as we do not become icebound we should reach our destination in two to three days.

It was –9°C, so I layered up before walking round to the stern to look at the long lane of freshly sliced, diced and pulverized ice that stretched without deviation to the horizon. The water was black. It reminded me of a slime trail left by a slug.

Back on the bridge, I saw that there were a couple of emperor penguins standing directly ahead of us. I wondered what they would do – and then I wondered what the skipper would do. Tall, imperious, unflinching, heads cocked to one side and watching us closely, they waited until the very last moment and then, when our bows were towering over them, tobogganed off, all dignity gone, printing their hooves in the snow as they went. The skipper threw me a look of relief. I asked how many penguins it would take before he would swing ship.

'Five emperors or eleven Adélies,' he replied without hesitation.

We were by then ploughing our way through an expanse that had thickened to almost a metre but, other than to dampen our speed, did little to hinder our passage. Then, suddenly, we broke into a patchy area of water on the far side of which was an immense sector of hummocked, multiyear ice that was two, three, and in places even four metres thick. In the far distance we could just make out another lead that seemed to be heading in roughly the right direction, but to reach it we would first have to wallop our way through the field ahead, which we could already tell was going to be obstinate. For a while we paused ship as the captain studied the front before us, looking for a line of least resistance. We were just 50 metres from the edge. Squared off as we were, it reminded me of a couple of boxers giving each other the big look before a fight.

Captain Freddie put down his binoculars, swung ship a few degrees and then opened the thrusters. The ship charged ahead, churning up the water beneath our stern. I watched as our great bluff bows went surfing up and over its edge, then came down upon it like a clenched fist. The ice shattered into chunks, and two great cracks opened before us.

We carried on like this for a while. It was tough going, and several times we had to charge more than once to achieve a breach. At present the brute determination of our ship to get through is greater than that of the pack to resist, but it would not take much to stop us. We have tremendous power when moving forward but if our way ahead were to become barred while we were in a tight channel of our own making, we probably wouldn't be able to turn around or move sideways. Then, as the pressure within the ice built when the tide fell and temperatures plummeted, the path in our wake could very quickly close, preventing any movement astern. At which point we would be in trouble.

I went back to bed. The first message I saw when I woke several hours later was from Claire Grogan, the team doctor:

Hi Folks. Just a reminder as we are heading into deep
ice . . . if you have any medical niggles, any small illness,
or even a vague sense of unease, please come and tell me
ASAP. We are going to be really far from definitive medical
help and it would be really difficult to get someone out. It
is easy to sort small problems while they're still small.
　Stoicism is dead.
　Thanks, Claire.

After breakfast, we had the daily management meeting.
Usually these are rather relaxed gatherings with cups of coffee
and a few laughs, but today nobody was smiling. It is now
serious. Frazer, the glacial geophysicist, began by taking us
through the most recent TerraSAR satellite images. There were
leads, but there were also vast areas of consolidated multiyear
ice. He ended by confirming something we had been thinking
for some time: although there are some stiff areas ahead, for
our purposes, this is the 'best' ice season since 2002. 'We have
been lucky,' he said.

At lunch, John went up on the bridge to confer with Captain
Knowledge and came back saying we were 80 nautical miles from
the search box. Although we were currently making good progress,
the last 40 nautical miles were going to be a slog.

Suddenly, as if out of nowhere, there were seals milling all
around us – lots and lots of them. We have, it seems, intruded
upon another world. Normally, nobody enters the central Weddell
Sea pack; there is no justification for putting yourself in such
danger. But this morning we kicked in the door and did just that,
and now we are witnessing things that few, if any, have seen since
Shackleton. We seem to have stumbled upon a peaceable Antarctic
kingdom, a secret, mystical wonderland where nature is truly
sovereign.

Strangely, it seems as if we are actually attracting the seals,
because every time we pause ship they start to gather in our wake
or haul up on the ice beside us. As the day has progressed, they

have been steadily growing in number. We have become some kind of Pied Piper for pinnipeds.

It might just be that we are driving deeper into their heartland, where they reside in greater profusion; but I am not so sure. There is one seal with a very distinctive scar on his neck from a tussle with some other species of fin-footed carnivore, and he has stayed with us for many hours. He obviously doesn't want to be left behind. If he is typical of the others, then we are gathering a retinue as we go. I grew up with seals. Seals are my friends. These seals have no fear of us, nor are they just tolerating our presence, as seals often do; and, since we are not feeding them, protecting them or creating opportunities for sex, the conclusion is that they simply want to be with us. In other words, they are curious about us and actually enjoying our company.

Other than occasional leopard seals and the even less frequent Ross seals (of which I have not seen any on this trip), there are just two kinds down here: Weddells and crabeaters. They are quite easy to distinguish. Weddell seals have catlike faces, with upturned mouths that seem to give them a smile, while the crabeaters (which do not, in fact, eat crabs) have pronounced and rather pointed snouts. On the ice, the latter are also easily identified because of their snakelike movements. Although we have seen some Weddells, the vast majority of those we are looking at are crabeaters.

NOON POSITION: 69° 19.945' S, 048° 57.198' W

Rifts within Shackleton's team. The search box.
We become icebound

Thomas Orde-Lees, or 'the Colonel', as he was mockingly called, is a case study in how easily relationships can sour during times of extreme stress.

Given their isolation, the exceptional menace they faced and the growing strains within the group, it was inevitable that some of the *Endurance* men slipped into a mood of despair. Even the irrepressible Worsley sometimes became dejected. One diarist commented: 'Worsley has lately been silent and morose, his conversation being chiefly abortive attempts at sarcasm. Today he put his foot into it badly by telling us that he "had given up talking nonsense" in reply to a query as to his sudden change from extreme talkativeness to almost morose silence . . .'

They were all on the rack and although some were more resilient than others, breakdowns were inevitable. In a letter to his wife from Port Stanley, Shackleton recalled that 'toward the end about 10 of the party were off their heads'.

Much of the despondency and ill feeling that occurred can, to some extent, be attributed to the gnawing hunger they experienced, which became worse as time progressed and their supplies dwindled. Again and again, food features in the diaries as both an obsession and a flashpoint. For instance, while on Elephant Island Hurley was accused of cheating on the food; it must have been a serious matter because even the wonderfully discreet and imperturbable Wordie took a swing at him when he wrote: 'There has been a change in the messes into which we are divided. Some of the No. 5 mess took exception to Hurley's acting as their server, which he was turning to his advantage. A rearrangement of the

1. The team of the Imperial Trans-Antarctic Expedition on the fo'c'sle deck of the *Endurance* soon after leaving the River Plate for South Georgia. Shackleton in white, in the centre behind the rail (1914).

2. The Weddell Sea Expedition team and the ship's crew on the pack ice beside the South African icebreaker S.A. *Agulhas II* (2019).

3. The Endurance22 team and the ship's crew on the helideck of the S.A. *Agulhas II* (2022).

4. S.A. *Agulhas II* in 2019 within what Shackleton called 'dense pack of a very obstinate nature'. The vessel cannot move forward or sideways and the freshly cut channel astern has closed and frozen over.

5. Sir Ernest Shackleton, 'the Boss', at the binnacle on the bridge deck of the *Endurance*.

6. Frank Worsley, the captain of the *Endurance*. His boatmanship and navigational skills were essential for the safe arrival of the *James Caird* at South Georgia, thus ensuring the eventual rescue of his shipmates left on Elephant Island.

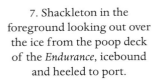

7. Shackleton in the foreground looking out over the ice from the poop deck of the *Endurance*, icebound and heeled to port.

8. The *Endurance* down in the ice soon after she was abandoned on 27 October 1915. The men are now alone on the ice at the centre of the most hostile sea on earth with no hope of rescue.

9. Frank Wild, expedition deputy leader, contemplating the wrecked upper decks of the *Endurance* shortly before she slipped beneath the ice.

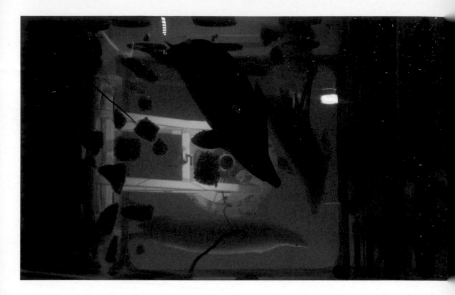

10. The moon pool on the S.A. *Agulhas II* provides sheltered access to the sea below for the safe deployment of small submersibles and scientific instruments. On 11 February 2019, the doors of the moon pool were opened and a number of crabeater seals swam up into the ship.

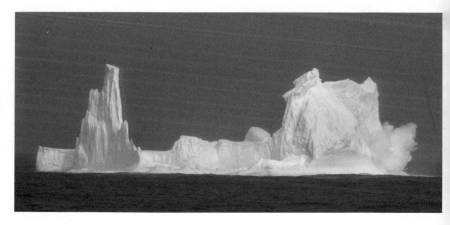

11. One of the great many icebergs that crowd the Weddell Sea. Their ice has been formed over thousands of years by the gradual freezing of snow and water on inland glaciers that eventually reach the sea, where great chunks carve off into bergs.

12. Professor Wolfgang Rack, one of the ice scientists, preparing to drill through a floe.

13. The main search-and-survey vehicle for the 2019 campaign was a Kongsberg HUGIN Autonomous Underwater Vehicle (AUV), here seen being lowered onto the helideck. Rated for depths of over 6,000 metres, it had a cruising speed of over 3 knots.

14. The *Endurance* trapped in the ice perilously close to a giant iceberg.

15. The *Endurance* in the ice with all square-sails set.

16. The *Endurance* seen through the rafted ice of a pressure ridge.

17. The cook of the *Endurance*, Charlie Green, in the galley preparing an emperor penguin for the pot. He was reputed to be the hardest-working member of the team. Both in the ship and on the ice, he had to feed twenty-eight men every day.

18. S.A. *Agulhas II* at Penguin Bukta, Queen Maud Land, with its bow pressed up against the edge of the Fimbul ice shelf. Teams and supplies that had been flown in to the Antarctic runway at Wolf's Fang were winched on board using the ship's crane.

19. During the Weddell Sea Expedition of 2019, football on the ice was popular – just as it was with Shackleton's men in 1914 (*above*). 20. *Below*: During one game, a group of curious Adélie penguins walked across the ice to where the game was being played, and stormed onto the pitch. The game ended. The author, in goal in the background, looks on bemused.

messes seemed the only solution.' On another occasion, while on the ice, the less forbearing Macklin pencilled a brief rant on how 'the cook suffers from bias and does not give fair whacks'. One man in Tent No. 5 recorded the following: 'This evening Blackborow who was tent Peggy [the one who collected their food from the galley] told us that the cook had served us a very small whack, giving the best share to his own tent. Lees took our whack to No. 1 tent and showed it to Sir E who agreed it was small and will see to it tomorrow.'

One can imagine that Green, the cook, would have bitterly resented being impugned before the Boss in this manner by Blackborow, who was his helper in the galley. The atmosphere between the two of them afterwards must have been caustic, but because neither kept a diary we will never know for certain.

Orde-Lees was the most observant and introspective of them all, and because of this we know more about his spats than those of anyone else on the team. There is no doubt that the relationship between the 'Colonel' and the Boss was complicated. At times Shackleton was kind to Orde-Lees, as when he nursed him through illness; but he did nothing to discourage the derogatory epithets and nicknames given to him by the team ('the Old Lady', 'the Belly Burglar', 'the Colonel', 'the Man of Action', etc.), which must have been hurtful. Nor did he hesitate to participate in the almost ritualistic humiliation of Orde-Lees, as, for instance, when he forbade him from going out onto the ice unless accompanied by Worsley. Sometimes one has the feeling that the wounding of Orde-Lees had almost become a sport.

Orde-Lees's great antagonist on the team was the grumpy Glaswegian Presbyterian 'Chippy' McNish, whom he described as an 'ill-mannered brute' with an 'exceptionally offensive manner', 'objectionable [and] cantankerous'. To Orde-Lees, Chippy was 'a perfect pig in every way'. In Chippy's diary there are several entries that make it clear these feelings were entirely reciprocated. He particularly enjoyed it when Orde-Lees, who was the motor mechanic, was shamed by having to depend on Hurley for the repair of his engines; and he was beside himself with joy when Orde-Lees was chased by a leopard seal and then, on another occasion, made to look foolish by losing his bicycle and then getting lost out on the ice.

Under the boats at Elephant Island Macklin found himself tight up against Orde-Lees at night, something which Macklin hated, not just for Orde-Lees's snoring but for his habit of 'jerking about very violently in his bag'. 'He is a terrible man to have next to one,' he concluded. Inevitably the sparks flew. Orde-Lees described how he inadvertently 'encroached upon Dr Macklin's allotment and as he suffered from incompatibility of temper, he had a good deal to say on this subject, which went in one ear and out the other like water off a duck's back'. The situation, however, was so explosive that the second engineer, Alexander Kerr, had to intervene and take Orde-Lees's allocated area in order to keep the peace. One of the cruellest jibes against Orde-Lees while on Elephant Island also came from Macklin when he wrote: 'We have long ago ceased to take him seriously.'

Another with whom Orde-Lees frequently clashed was Worsley, but so too did just about everyone at one time or another. On the voyage out from England Worsley managed to irritate most of the ship's company by his rather whimsical style of command, to the extent that Shackleton had to appoint himself captain of the *Endurance* when he joined the ship in the River Plate. In particular Worsley did not get on with McNish – which may in part explain why the old carpenter was later denied the Polar Medal, for Shackleton would have sought the skipper's advice on the matter. Most prominent, however, were Worsley's differences with Orde-Lees. '[Orde]-Lees and Worsley do nothing but argue,' wrote one diarist. With typical understatement, Orde-Lees described an occasion in the tent with 'Wuzzles' when, 'Words began to pass between us, and the situation became a little strained . . .'

One of the most melancholy episodes to appear in the diaries happened when, in effect, Orde-Lees was driven from his tent to live alone in the makeshift pantry they had erected at Ocean Camp.

No doubt there was collusion, but it was probably Worsley who led the push, as he was tent leader. On 10 November 1915, Orde-Lees wrote: 'There is a movement afoot to eject me from the eight-man pole tent and make me sleep in the rabbit hutch.' The following day he added: 'the occupants of the eight-man pole tent have successfully evicted me to the rabbit hutch'. To have been so cruelly rejected by his tentmates must have been painful for the sensitively wired marine captain. The day after they gave him the boot, in words dripping with sarcasm, Worsley recorded the following for posterity: 'Sounds of bitter sobs and lamentations are heard this evening from No. 5 tent at the loss of their dearly beloved "Colonel" who has removed himself for a season to sleep in his store in the old wheelhouse. He indulgently yields to our earnest entreaties to continue to dine with us and comforts us with the assurance that he will return promptly to our humble but happy home immediately we prepare to get on the march.'

The cracks that had first opened up within the team at Patience Camp were much more pronounced at Elephant Island, and the rift between Wild and Orde-Lees became truly acrimonious. Wild had been left in charge when Shackleton sailed for South Georgia in the *James Caird*, so there was nobody to settle disputes between the two when they arose, which was often. Their arguments at first were over whether or not they should kill more penguins, after which, in Orde-Lees's words, 'an estrangement' opened up between him and Wild and soon turned ugly. Wild should have applied his authority in a more sensitive manner when dealing with the troubled Colonel, whose line of reasoning, in hindsight, was often the wiser; instead, he used his position to bully and inflict derision. One of the diarists gleefully relayed how Wild 'found it necessary to coalhaul a certain member (L[ees]) for shirking certain night duties'. On another occasion he cruelly pilloried the hapless Orde-Lees for his snoring by 'making him tie a noose round his arm'. Jimmy James described how 'The rope passed across my bed and Hurley's and through two eyelets across to Wild and McIlroy. When the snores resound a pull on the rope is supposed to waken him.'

Orde-Lees wrote of this rather sadly in his diary: 'Of course everyone thinks I snore deliberately to annoy them, but then everyone here always does think that everyone else does do everything nasty deliberately. Occasionally Wild loses his patience, waxes wrath and pulls the cord unmercifully swearing volubly.'

Most would agree that Wild did extremely well at holding things together on Elephant Island. Nonetheless it does seem that in the absence of Shackleton, team unity was beginning to unravel, as evinced by the theft of provisions – something that probably would not have happened had the Boss still been there. The finger of suspicion was hovering over the fo'c'sle hands, or 'the men of the lower deck' as they were sometimes called.

Relations with the fo'c'sle hands further deteriorated when the team's tobacco supplies began to dwindle. Anticipating trouble, Wild had the remaining stock distributed equally among the smokers. The sailors, however, raced through their rations without seeking to make them last and then, when they ran out, became resentful of those who had been more prudent.

Perhaps the most explosive moment in the entire Shackleton story was a fistfight that occurred at Elephant Island between one of the sailors and Hurley. Although the seaman was not identified, it was most likely Vincent, the former bosun who had been demoted by Shackleton for bullying his fellow fo'c'sle hands.

Of course, none of this accords with the picture of the 28 heroes that has passed into folklore. It must be remembered that Shackleton's men were just blood and bone, and long, arduous expeditions always draw forth the worst as well as the best in people.

Last night it seemed as if, providing we could sustain it, the rate at which we were slamming through the pack might put us over Worsley's famous coordinates by mid-morning.

Next to the cabin where I sleep, I have a rather luxurious day room with sofas, tables, desks and large square ports looking out over both the bow and the starboard beam. For a long time before going to bed, I just stood there watching the sea ice folding beneath our onslaught into great slabs, which then passed down the sides of the vessel before tumbling into our wake. The violence was awesome. Everything shook from the impact; the trauma experienced by the vessel was almost as great as that inflicted upon the ice. With every charge, the wood panelling in my quarters creaked, and there was a chorus of other straining sounds from unseen elements deep within the ship's anatomy. Mingled with that was the noise from the rupturing pack outside.

Shackleton had with him Coleridge's 'The Rime of the Ancient Mariner', so of course I also had to bring a copy with me:

> The ice was here, the ice was there,
> The ice was all around,
> It crack'd and growl'd, and roar'd and howl'd
> Like noises in a swound!

How, I wondered, did Coleridge ever write stuff like this when, the story goes, he hadn't even crossed the Channel?

I finally got to bed at about 0100. For a while I just lay there, feeling as if I was in some kind of alternate universe. I'm not sure how long I slept before I awoke with a slight sense of anxiety – something had changed. There was silence. Total and absolute. None of the woodwork was complaining, nor were any deeper protests arising from within the hull. In fact, there was no movement whatsoever: no motion, no heaving, and certainly no icebreaking. Just nothing.

I got up and went into my day room. From the port looking out over the fore deck, I could see that we were totally stationary: encased on all sides by a great field of ice. Although there was fog, I could just make out a couple of feeble leads to starboard, but nothing close and certainly nothing heading off in our direction.

We were stuck fast.

Macklin talked of the 'dead Antarctic stillness'. Other diarists, and indeed Shackleton himself, spoke of the profound silence that spread over everything when the wind dropped. I went out on the side deck, and for the first time I understood what they meant. There was no sound from the ship, no movement around us, no wind, just complete quietude.

I climbed the steps to the bridge. The mate was standing over the ECDIS, the electronic navigation display.

'We icebound?'

He looked a bit stressed. 'No,' he snapped, 'we're just nipped. As soon as the visibility clears we'll be on our way.'

And that was the message that went out. The crew, however, were snorting with laughter at this. 'We're as fast as a cork in a fuckin' bottle,' I heard the bosun's mate say to one of the South African students.

I think the reason this is being played down is that, understandably, they do not want to unsettle anyone. Later I talked to Captain Knowledge and he said he was waiting for the tide to rise, which always loosens the pack. Soon afterwards I spoke to Captain Freddie, the ice skipper. He knew I was concerned about the loss of time and what it could mean for our search. 'Don't worry, Mensun,' he said, 'I'll get you there. I still have a trick or two up my sleeve.' Really? I thought.

By lunchtime the story was wearing a little thin and everybody was talking about being beset. There were even nervous jokes about having to spend the winter in the pack, like Shackleton.

It wasn't until early afternoon that I sensed a little movement. Then, going to my window, I saw what Freddie had meant by a 'trick or two'. The crane had picked up a 16-ton fuel container and was moving it from one side of the ship to the other, creating a rocking movement that was intended to rupture the ice at its point of grip.

This is an old technique called 'sallying' ship, often used when a hull became stuck in the mud and could not move. To break the suction, they would attempt to roll the hull back and forth by having the ship's company run in unison from one side of the weather deck to the other, then back again. The same approach had even been used by Shackleton in an attempt to release the *Endurance* when she became fast in the ice. As he described it: 'all hands joined in to "Sally" ship. The dog-kennels amidships made it necessary for the people to gather aft, where they rushed from side to side in a mass in the confined space around the wheel. This was a ludicrous affair, the men falling over one another amid shouts of laughter without producing much effect on the ship. She remained fast . . .'

Unfortunately, it didn't seem to be working much better for us. At about 1600 hours word went out that we should be on our way soon and indeed, for a very short distance, we were – but then, once more, the pack took us back into its icy embrace. This time there was no more talk of waiting for visibility to improve. We were truly stuck and all we could do was wait for the next high tide. And, we were told, hope.

It is hard to convey the intensity of the frustration that welled up inside me at that point, as we sat clamped for the second time. Figuratively speaking we have journeyed 10,000 leagues, and in my dreams much further. We are so close now, I can almost reach out and touch the spot where Captain Worsley stood as he watched his ship disappear through the ice.

In normal circumstances, one might have expected the mood at dinner this evening to be sombre. But not at all. In large part this can be attributed to something golden that happened during the afternoon.

Yesterday I spoke of how we were being swarmed by crabeaters and speculated that they were increasing in number, and that they seemed in some way to be attracted to us. Of this there can no longer be any doubt. In growing profusion they have been disporting themselves in the open water and frazil off our stern. When they want a rest, they haul up onto the pack and go to sleep.

Soon after we became fast for the second time John reported to the team that from the bridge he had counted over 100 seals on the ice beside the bow, and as many more again in the small pools around the ship. About an hour later, I made it 135 seals, and by mid-evening their population was still expanding.

Everybody recognized this as an extraordinary moment that we will probably never experience again in our lives, so we bundled up and spent as much time on deck as the severe cold would allow. At one point Ray yelled for me to come over and see something. I joined him at the gunwale but at first I couldn't make out what he meant. Then I heard an escape of air, as if somebody had just stabbed an overinflated football. Looking down, I could make out a number of seals suspended, motionless, in a vertical position, with only their nostrils protruding through the brash ice beside the ship. Off our starboard bow there was an open pool – looking into that, I could see it was full of crabeaters too. Most remarkably of all, they were actually coming right up to our submerged hull and nuzzling us with their noses.

We discussed why they were doing this. Could it be that we were radiating a little heat? Or was it the vibration – or maybe they were just exploring the novel texture of steel? One thing was certain: they wanted to be here. They were free to go anywhere, but they were deliberately choosing to club up beside us. There were those among us who spoke as if there had been some kind of crossing of the divide; as if there had been a fragile moment of mind-touch between our species.

There is a long tradition in literature, poetry and even music, of other worlds. Invariably mysterious and inaccessible, sometimes where time stands still, but always where there are marvels to behold and strange things occurring. Think of Hilton's *Lost Horizon* or Edgar Rice Burroughs's *The Land that Time Forgot*, neither of which I have read; or of Jules Verne's *Mysterious Island*, Rider Haggard's *King Solomon's Mines* or Conan Doyle's *Lost World*, all of which I have. And then there is Edgar Allan Poe's 'Narrative of Arthur Gordon Pym', which, mercifully, has been largely forgotten, but which actually features a hidden land here in the Weddell Sea that is home to an unfriendly indigenous population. Poe's story was inspired by the seafaring exploits of Benjamin Morrell, an American sea captain who in 1832 published an account of one of his voyages in which he entered the waters where we are now. He called his 'discovery' New South Greenland. In time, it became better known as Morrell's Land. Its existence was doubted, but it was not until Wilhelm Filchner's *Deutschland* expedition of 1912, and more especially the drift of the *Endurance* three years later, that its existence was disproved to everybody's satisfaction.

All of these hidden wonderlands were fictional – the fantastical imaginings of storytellers – whereas this icy arcadia into which we have intruded, and the strange things we have witnessed today, are very real.

Everything now depends on the next high tide. If that can lessen the squeeze, there is a chance that we might be able to elbow our way out of here. And if we manage that, by tomorrow afternoon, AUV 7 could be down and sniffing around in the shadows.

NOON POSITION: 68° 57.528' S, 052° 04.494' W

10 FEBRUARY 2019

We reach Worsley's coordinates for the sinking of the Endurance.
AUV 7 dives. When the Endurance *began her plunge*

We were the first that ever burst
Into that silent sea.

—SAMUEL TAYLOR COLERIDGE,
'THE RIME OF THE ANCIENT MARINER'

At 36 minutes past midnight, we broke free of our icy fetters and the *Agulhas II* slowly hauled herself back to life. The sun, low and crimson, left the ice glowing like a flambéed crêpe suzette. There were then just 19 nautical miles between us and our destination, but we knew that last little bit was going to be both tortuous and torturing.

At this point, those of us still up retired to bed, not expecting to reach the search box until breakfast. I went out on the deck for a few seconds before the −13°C temperature drove me back inside. By then the ship was on the move again and breaking through the ice pack.

I have been reading James's unpublished diaries. In his entry for 4 April 1916, there is a line that jumped straight out at me. The men had been carried on the gyre to the fringes of the pack and now the ice was literally breaking up beneath them. The incident in which Ernie Holness, still in his sleeping bag, had been dropped into the sea was fresh in their minds. After more than five months on the ice, it could now only be a matter of hours before they would have to take to their boats in an effort to make land. James wrote: 'We are on the verge of something, there is no doubt.'

And that is how I feel. We are almost at journey's end. It feels like my whole life has been narrowing to this moment.

At 0524 hours, the ship's tannoy crackled to life. It was the ice skipper, Freddie Ligthelm: 'Good morning from the bridge. This is to say we have reached the *Endurance* sinking position. *Lekker, lekker, lekker* [Afrikaans for "Nice, nice, nice"]!'

In an instant I was out of my bunk, pulling on my clothes and racing for the bridge. I expected a cavalcade of people, but I was the only one. On the bridge were Freddie, another officer, a deckhand and one of the students on ice monitoring duties. Freddie and I shook hands solemnly and I thanked him.

There was an aura to the moment; we both felt it. Freddie was one of the design team behind the *Agulhas II* and oversaw her construction. Naturally he is pleased for the ship, and more than delighted that she has reached this sacred spot in polar history with him at the helm. As for me, with the vessel on station over those famous coordinates, I feel the presence of Shackleton himself as never before. For the brazen little *Endurance*, this is where her voyage of no return ended.

There was no wind and, although it was overcast, visibility was good. A glance at the depth sounder told me that we had 3,038 metres of stygian black ocean beneath our keel. There was a tight mosaic of multiyear ice all around us, but – and this was vital – there were also pools and a couple of minor leads, some of which were large enough to launch an AUV.

Anticipating this moment, Freddie had instructed the crew to prepare a wooden sign commemorating the achievement. It reads:

S.A. AGULHAS II

ENDURANCE

WRECK SITE

68° 39' S, 052° 26' W

He and I signed it; throughout the day, everybody on board would be invited to add their signatures. As he put down the pen, Freddie looked at me with a slightly triumphant expression. 'I got you there,' he said. Stout fellow, Freddie.

The next up on the bridge was John. There followed a stilted moment when we all looked at each other as if to say, 'What now?' For days we have been laser-like in our determination to reach this spot, but now we have arrived – and have done so slightly ahead of schedule – we suddenly find ourselves caught off guard. We could have scrambled the team, but we knew that they had a very long day ahead of them. From this moment on, sleep was going to be at a premium and on an opportunity basis. We decided to let them have their last hour of rest before breakfast.

Freddie suggested I grab another hour of sleep myself, but from his lopsided grin I knew he wasn't serious. It reminded me a bit of the Apollo 11 landing. I once read somewhere that, in the original plan, once the *Eagle* had settled on the lunar surface, Neil Armstrong and Buzz Aldrin were supposed to have several hours of shut-eye before undertaking their epic moonwalk. Who, I wondered, had dreamed that one up? You are about to become the very first humans to set foot on another astronomical world, and someone in Houston tells you to take a nap!

So, until breakfast, I just stood there.

Both John and I have brought with us embroidered copies of Shackleton's sledge flag from the *Nimrod* expedition. Sledge flags are a British custom believed to have been started by Captain Horatio Austin, who, in 1850, led a naval search for Sir John Franklin's lost expedition of 1845 to the Northwest Passage in the Canadian Arctic. To encourage *esprit de corps*, he gave each of his sledge teams a distinctive flag to take with them on their journeys across the ice.

Shackleton's sledge flag was yellow, with a central horizontal red stripe charged with three lozenge-shaped buckles. In the top corner was a red canton featuring a cross. Until now our copies of this flag have hung above our desks, but today we decided we would entrust them to AUV 7 for her voyage of discovery, sealing them within the fuselage. It was our little nod to history.

A short time ago I was made a Fellow of the Explorers Club of New York and they approached me about being their expedition flag-bearer. This is a particular honour, as Shackleton was one of the most eminent members of this prestigious society and, therefore, the hunt for the *Endurance* is a matter of huge interest to the Club. In addition to my Shackleton flag I have recently been thinking about sending the Explorers Club's flag down with AUV 7; I even discussed the idea with Julianne Chase, a vice president of the institution. It wasn't as straightforward as it sounds because the Club's flags are carried on many expeditions, so each is imbued with a history of its own that goes back many decades. The one I carry is of particular value to the Club.

We both dithered. Finally I put it to her point blank. 'Julie,' I said, 'it's your call.'

'Do it,' she gulped.

Today, however, at the very last minute and for reasons I can't quite explain, I changed my mind and did not send the Explorers Club flag down with the AUV.

Before we could launch AUV 7 we had to carry out a conductivity, temperature and depth cast so that the AUV's systems could be effectively calibrated. While this was happening the Ocean Infinity guys were deep under the hood, conducting their final audits of the vehicle's technical innards. All the main sensor arrays were powered up to ensure 'full operational compliance'. At the same time the propulsion unit was tested, along with the rudder and all other mechanical components. Also, the submersible's inertial navigation functions were put through a review process so that they could all be confirmed performance-ready. Finally – just in case the AUV aborted its mission and made an unplanned ascent and the ship (which could be many miles away) had to execute an emergency recovery – trials were undertaken to ensure that the vehicle's GPS and Iridium satellite habiliments were in 'full communicative state'. Normally, for a launch, the ship turns into the wind and holds speed at two to three knots, but today – and here we were very fortunate – there was calm, enabling the ship to throttle back to 'slow ahead' with her bows pointing into a large patch of open water.

By 1220 hours all the pre-dive checks had been completed and everything was 'Go to commence launch.' Blake Howard, who was in charge of the ops room, informed Devon James in the hangar on the back deck that AUV 7 was 'in armed state' – ready to dive. Devon then radioed the AUV team leader, Channing Thomas, on the bridge.

The hydraulic launch and recovery cradle, or 'stinger', was extended out over the stern and hinged down so that AUV 7 was suspended over the water at an angle of 45 degrees. Once Channing was content that the ship was on the right heading and speed, and that no evasive ice manoeuvres were in prospect, he gave the instruction to proceed with launch, at which point responsibility reverted to the back-deck team. Devon took one last look around to make sure everything and everybody was mission-ready, and then shouted 'Pull!' Todd Oxner immediately whipped back on the magnetic start key that energized the submersible and, in the same instant, AUV 7 began to slide, rear-end first, down the chute and into the sea, where Chad Bonin was waiting with his team in a fast rescue craft in case anything went amiss.

It all sounds rather complicated, and it was. But because they are all highly trained specialists, everything proceeded seamlessly.

At 1234 hours, to a ripple of applause from the assembled team on the helicopter deck, AUV 7 began its history-making dive under the pack and down into the deepest recesses of the Weddell Sea. Normally AUVs leave the surface at an angle of 20 degrees, but because of buoyancy concerns and the close proximity of the ice, ours was put on a steep downward trajectory of 45 degrees. We kept it in supervised mode throughout the descent and then, when it reached loiter depth, there were the usual position updates and systems checks before it was released to begin its under-ice transit to the search box. At 1339 hours, Blake reported to the bridge that AUV 7 was 'at operational depth and on task'.

The only difference between this launch and the countless others

I have attended was that a five-kilo paper bag of South African table salt had been taped to the upper fuselage to give the AUV the extra weight it needed to get its nose under quickly. This was the idea of Espen Strange from Kongsberg, the Norwegian manufacturers of the AUV. It's a technique he has employed before during AUV trials in the Baltic (although on those occasions he used flour). What happens is that once the vehicle has submerged, the paper bag quickly turns to mush and disperses its contents into the water, thus allowing the AUV to re-establish equilibrium before going into cruise mode. I argued today that we should use sugar, which would dissolve rather than wash away; however, sugar was in limited supply. But there was plenty of salt, which in any case was thought to be more environmentally friendly.

And so AUV 7 is on its way. After days of working on flatscreens and with coordinates on paper, my life has suddenly gone three-dimensional. I like it.

As soon as AUV 7 was released from its first loiter station, it set off at 3.6 knots on a two-hour under-ice transit to the southeast corner of the search box. On arrival it will again go into a holding pattern while final checks are conducted, after which it will resume autonomous running and commence the search. Once on mission it will follow a path of programmed coordinates along a series of parallel lines; basically connecting the dots as it goes. Periodically it will reach prescribed rendezvous stations, where there will be what we call a 'handshake': those in mission control, or what is sometimes known as the online room, will check its sensors and navigational systems and correct any deviations or other irregularities, the most likely being drift. When the operator in charge is satisfied, the AUV will be released to continue the survey.

We are not operating in real time, nor do we actually 'see' the ocean floor in the normal sense of the word. Yes, we are searching, but as if with bandaged eyes. There are a number of payload systems within the AUV's fuselage but the most relevant is the side-scan sonar that emits a downward, fan-shaped spread of sonar pulses, which are reflected back to the sensors by the landscape within its swathe. This gives a series of 'slices' across the terrain perpendicular to the direction of travel, and these are then digitally stitched together to produce a picture of the seabed. Because of the way the side-scans are orientated there is a gap directly under the vehicle, and this is filled with sensor readings from a multibeam echo sounder that generates both a 3D bathymetric image of the ocean floor and a 2D display, similar to that of the side-scan sono-gram. The side-scan method is often compared to entering a dark room with a torch: the faces of objects caught within its beam will be highlighted, while the area behind them will be bathed in shadow. In principle, this is similar to what we are doing, but instead of using light we are using sound.

The altitude (height above seabed) at which we fly the AUV can be critical. Fundamentally, this is determined by the nature of the target as well as that of the seabed. If we operate the side-scan from a high altitude (i.e. at a low frequency) it means we can cover a wide area quickly, but resolution will be poor. By contrast, if we go in low (i.e. at a high frequency) our field of vision will be narrower, but we will be able to see what's there in greater detail. Determining the right altitude for the task at hand is not always easy.

Unlike ROVs, which give us a real-time, fish-eye view of what is there, an AUV is more like a message in a bottle: it takes time. You have to wait until the voyage is over before the payload data can be transferred to the server for conversion into a format that can then go to the processors in the offline room for what we call PMA, or post-mission analysis. The downloading of the raw payload data requires about two hours, but the interrogation of the results that follows can take a further three hours or more, depending on the complexity of what has been gathered by the AUV. In other words, it can take over five hours before we know whether the AUV has encountered anything interesting or anomalous. I have spent great chunks of my life working with side-scans and can therefore read them as well as the specialists, so when the data is released from the server, I will be there with the analysts. Whatever the nature of the mission, it's exciting to be there as the results come through. That's always the bit I enjoy most.

So what happens if AUV 7 locates the *Endurance*? First, I know that it is extremely unlikely that she will look like a ship. She will most probably appear in the sonograms as nothing more than an incongruous shape with an array of shadow-throws along one side; or worse, if she is broken up, then as a number of smaller abstract forms reminiscent of rocks. Being able to discriminate between geology and wreck is where skill and experience comes in. You can have the most advanced kit and the best technicians in the world, but all that will count for nothing if you do not have a skilled team of analysts to interpret the findings.

On a featureless seabed any man-made deposit of wreck size will stand out, but I suspect the bottom of the Weddell Sea will be characterized by numerous drop-stones from icebergs, and this could complicate matters for us. If, however, AUV 7 identifies an anomaly, or what we call a POI (point of interest), we will dispatch AUV 9 on a reconnaissance mission to confirm whether or not it is the *Endurance*.

AUV 9 was specially designed to study the underside of ice shelves; its multibeam echo sounder cannot scrutinize the seabed, it can only look upwards. It does, however, contain other highly advanced systems including a still photography unit which, when skimming the bottom at less than 10 metres, can capture everything in tiny detail. In addition, from further above the seabed, it can conduct a so-call 'high-resolution interferometric synthetic aperture sonar survey': HiSAS for short. This bit of gear comprises an array of sophisticated transmitters and receivers along either side of the vehicle's fuselage that gather and combine various data sets to give a detailed bathymetric account of the seabed as well as high-resolution imagery of any objects within that landscape. Originally this system was designed for the military, for the detection of seabed mines and listening devices, but it's now proving an outstandingly useful tool for maritime archaeologists.

While AUV 9 is on mission, AUV 7 will have its batteries exchanged and go through the usual pre-dive checks in preparation for redeployment. Normally when we are on search-and-survey missions we dispatch multiple AUVs at once, but because of the challenges associated with the pack, it was decided that there could be no 'sim ops' (simultaneous operations) this time. We must wait until 9 is back and then, if the target is confirmed as the *Endurance*, we will launch 7 again to conduct the final recording dive, which will focus mainly on image-gathering for 3D modelling. This will be my main source of information for archaeological analysis of the wreck.

But for the moment, AUV 7 is down there in the deep, dark, unlovely maw of the Weddell Sea, searching for the most famous wreck of them all. Its mission will require 42 hours of bottom time. All we can do is wait and wonder. Will it be an incandescent moment of discovery, or a kick in the teeth?

On what the men called the 'fateful day' – Wednesday, 27 October 1915 – Shackleton said of the *Endurance*: 'This is the end of the poor old ship. She's done for. We shall have to abandon her.' And they did, but she did not sink straight away. For another 25 days, although full of water, she remained pinioned to the surface.

They erected their tents just a short distance from the wreck (Dump Camp), but less than a week later they re-established themselves at a spot between one and two miles away. It was from there (Ocean Camp), during the late afternoon of Sunday, 21 November, that they all watched as the *Endurance* made her final plunge. Shackleton, who was outside and looking towards the ship, suddenly 'saw the funnel dip behind a hummock'. He ran to the raised wooden lookout they had built. 'She's going, boys,' he shouted.

Worsley wrote in his log: 'At 4.50 p.m., Sir Ernest on the floe sees her funnel moving downwards and hails me in No. 5 tent. Without hearing what he said I somehow knew she was going. I rushed out and up the lookout where we watched the death of the ship that had carried us so far and so well and then put up the bravest fight that ever a ship had fought before yielding, crushed by the remorseless pack.'

Wordie wrote: 'we all rushed out and posted ourselves where best we could see her a mile away. The stern was right up until she almost made an angle of 45°; but her head was going down all the time and about 10 minutes after the first alarm, the last of her stern sank from view.' Understandably, the mood was subdued: 'No one said much. It seemed as if the moment of severance from many cherished associations, many happy moments, even stirring incidents, had come as she silently up-ended to find a last resting place beneath the ice on which we now stand.'

NOON POSITION: 68° 38.377' S, 052° 29.280' W

11 FEBRUARY 2019

Optimism brimming. Seals in the moon pool. Disaster

Overnight, four of the 11 lines were completed and we conducted three successful handshakes. AUV 7 was behaving well and all its payload systems were functioning. Tomorrow it will be back. Optimism is ballooning. In the inimitable words of Chad Bonin, 'If we can't do it, then fuck it – it can't be done.'

The mood on board is a bit floaty, almost festive. Group photos are being organized on the afterdeck, yoga sessions are taking place in the gym, badminton and table tennis are being played in the helicopter hangar, guitars can be heard in the lounge and small groups are busy making short-story videos on their phones for the 'Agulhas Film Festival' that has been organized by the documentary team. One enthusiastic young man with a clipboard tried to sign me up for a cricket match, to be held this evening in Hold 3.

To add to the feel-good mood, early this morning we received some excellent news from head office in London. Since reaching the ice edge, John and I have been deeply concerned regarding our deadline for when we must cease all operations and return to Penguin Bukta. The lost days at King George Island mean that if we do find the *Endurance*, we won't have enough time to conduct the necessary follow-up inspection dives. We have previously been told that there's no flexibility in the ship's schedule but somehow London has now managed to negotiate an extra 24 hours into our charter. Just like at Woodstock in 1969, we have stretched three days into four.

And so we have cast our harpoon. It's now not so much a matter of 'will it hit its target', but rather, is its target there to be hit? The question of whether the *Endurance* is actually in the search box has been eating away at me for days, but this morning I made a conscious effort to put aside my anxiety.

Of course, I still have my concerns, and the main one right now is the recovery of AUV 7. We have to locate a pool into which it can execute a controlled ascent, and at the moment the pack around us is so tight it will be like a little sprig of chamomile trying to find a chink in the paving through which it can wriggle its way to the radiance of the world above.

As for the ship, she's been smashing her way back and forth within the search box in an effort to reach rendezvous coordinates that will allow us to communicate with AUV 7. It is a truly challenging set of dynamics. As a rule, we do not deliberately seek to fight the ice; we try to work with it, or even allow it to impose its will on us; but now the balance has shifted and our relationship with the pack has become confrontational. The flight path of the AUV is locked to a small patch of unmoving seabed while we, 3,000 metres above, have to reach, and then for a while cling to, defined 'handshake' points along its route. The trouble is, we're now within the hard core of the Weddell Sea gyre and while we have to remain within the box, the pack, moving at two knots or more, is applying all of its gargantuan strength to driving us out.

With winter a spit away, our abiding fear is of becoming permanently icebound. We are beyond the reach of assistance so it is imperative that we do not allow ourselves to become a permanent ward of the pack. Clearly, the best thing for us is to be always on the move. Even during those short periods when we are on our rendezvous coordinates and waiting for the arrival of the AUV, we keep stirring the pot to prevent the brash and broken bits from congealing into a solid cake about us. This doesn't always work, and since the launch of 7 there have been some dry-throat moments when we were wedged in completely. One of the crew told me today that we have about five weeks of food left, after which 'we'll be noshing on seal just like Shackleton'.

John reviewed the protocols that would be followed should we find the wreck. Once I've confirmed its identity, all communications on the *Agulhas II* will be closed down except to the vessel's master; the expedition leader, John Shears; and myself as director of exploration. The master will notify the ship owners, i.e. the South African government; John will report to the Polar Regions Department of the British Government's Foreign and Commonwealth Office; and I will inform project HQ in London. I will also call Donald Lamont of the Expedition Advisory Committee so that he can brief the Shackleton family, who have been so supportive of the project. All comms on the ship will remain in shutdown until a press statement has been released by Celicourt, the project's PR voice in London.

And then there's my wife, Jo. I do not want her learning about it from a news flash, or a phone call from a well-meaning colleague. She has lived this dream with me. I want her to be the first to know that the *Endurance* has been found – and I want to be the one to tell her. Between me confirming the identity of the wreck and the closing down of all private ship-to-shore communications, there will be a gap of seconds: just enough time to tap out a few letters of text on my phone. So we've devised a one-word code that says it all. If that word is sent without an exclamation mark, it means that we have found the *Endurance*, but that she is broken up. If there is one exclamation mark, it means the hull is semi-intact and looking good. And if the word is in caps with double exclamation marks, then it means that she is knock-your-socks-off, eyeballs-out-on-their-stalks, gob-smackingly spectacular.

And what is that magic word? *Bingo*.

Earlier something very special happened that will stay with me for the rest of my life.

The skipper thinks there are now well over 250 seals in our vicinity. They crowd about us in the water and then, the moment we are icebound, they come and nuzzle the ship with their nostrils or haul up close beside us on the ice, where they either just lie there twitching like still-warm cadavers or roll over, tummies in the air, like well-fed pashas waiting for servants to dangle grapes into their mouths.

The two people on the ship who most share my passion for the wildlife of these high altitudes are Claire Grogan and Holly Ewart. They are always on the lookout for penguins, seals and whales and as soon as anything exciting appears, they notify the team. In the afternoon Claire sent out a team message to say '10 emperors swimming along port side,' to which Holly came back, 'Also about a billion seals.' Not long after that, I just happened to be crossing the science deck and heard a sound which told me they were about to open the moon pool, the yawning square shaft that goes right through the centre of the ship and allows us, when we are in ice-infested waters, to lower the HiPAP, the metal shaft with a transducer on the end that enables acoustic communication with the AUV. I have always been fascinated by life in the sea and so I love to watch when the moon pool trap, which is flush with the bottom of our hull, opens and the sea gurgles up inside the ship. Sometimes I am rewarded with a fleck or two of plankton.

This time, I could not believe my eyes. Looking down through the ship and out through the 2.4-by-2.4-metre proscenium into the depths below, I could see a frenzy of seals darting back and forth immediately below our keel. As I watched, they actually swam right up into the ship. And then, in a twinkle of immaculate magic, first one, then another, then yet another, broke the surface and stared up at us through big, dreamy brown bovine eyes, double-size.

Were I to wake up tomorrow on Alpha Centauri, there would not be wonders to match that moment.

If the experience of the seals coming up into our ship was one of the most sun-burstingly sublime moments of my life, what has happened since then will undoubtedly be remembered for the remainder of my days as the most pulverizingly awful. The instant when everything came to a screeching halt.

To explain fully what happened, I need to return to yesterday. AUV 7 entered the water at 1234 hours. Because of the proximity of the ice, it performed a steep dive to operational depth, where communications were established; then, following the usual checks, it began the two-hour under-ice transit that would take it to the search box. At 1425 hours we went crashing after it, so as to be poised on station to receive the signals when it came into the designated rendezvous area at the southeast corner of the search box.

We were no more than 40 minutes into our passage when, in the middle of a blasting gale, we became wedged between two vast floes. The pressure was such that the lane we had cut through the ice closed behind us like the stone door of a rock-cut tomb. The *Agulhas* was on full-ahead, but we were paralysed. We could move neither ahead nor astern. Great forces were at work and we were too puny to resist. In response to the gnarring of the ice, straining sounds rose from the body of the ship. Had we been made of wood like the *Endurance*, we would have been squashed like a bug. 'Fuck, fuck, fuck,' said the mate.

This was, potentially, very serious. We had to make contact with AUV 7 soon for a number of reasons, one of which was drift. We know very little about the bottom currents in the Weddell Sea, and the long transit that the AUV had to make in order to reach the search box could result in excessive course deviation. Ordinarily, with regular 'handshakes', this would not matter because there is enough overlap between the lines to absorb a margin of drift, and at each handshake any deviation is corrected. If, however, we start missing handshakes, drift will accumulate and significant gaps will open up in our coverage. Normally with gaps we return later to conduct what we call backfilling operations – but given the time constraints under which we are now working, this cannot happen.

Matters became tense but at last, as the tide reached its peak, we broke free. We had by then, however, missed our handshake with AUV 7. In such a situation the vehicle goes into a holding pattern within the station, which is a notional 200-by-200-metre box centred on the rendezvous coordinates. If we have not made contact by the time it has completed seven circuits of the perimeter, it will continue its mission. And this, indeed, is what had happened. It was now absolutely imperative that we arrive at the next rendez-vous point ahead of AUV 7 so, as soon as we were free of our icy fetters, we charged off, tusks down, like a bull elephant in pursuit of a safari Jeep.

To our great relief, we arrived on station with even a bit of time to spare. As soon as we were over the coordinates, we opened the trap on the bottom of the ship. The sea gurgled up into the moon pool and the HiPAP was lowered to below the keel, where it began to listen for pings. The AUV's inertial navigational system had been programmed to follow a series of parallel lines, the standard 'lawnmowing' configuration. We were by then sitting directly over line two. As long as nothing had gone wrong, AUV 7 should now come to us. Grim-faced, we waited in silence, all eyes fixed on the screens.

At 2155 hours the monitors suddenly lit up and AUV's projected navigation path went from purple to green, meaning that the Acoustic Command Link had been activated and we were in real-time data-exchange mode with the vehicle. For what felt like an age we had been rigid with tension, but in that instant a megawatt surge of elation passed through us all. Our project was back on track. Chad was at the console; we threw each other a glance. We had been in some tight corners before, but this was one of the worst. There were no whoops and nobody punched the air in jubilation; there was a job to do. Position updates were already streaming in.

Not surprisingly, AUV 7 had 'drifted' some 26 metres from the programmed path, but this was within acceptable limits. All payload systems were reviewed; all registered normal. The necessary repositioning correctives were transmitted to the vehicle's guidance system and it was released to continue the search.

The next handshake took place just before midnight, when AUV 7 was on its third line. This also went well. A slightly low battery temperature was recorded, but this was a routine issue of little concern.

And so, as we came into today, everything was once more looking good. A little after midnight, I passed through the lounge. There were still a few up, playing board games and listening to music. They wanted to know what was going on in the ops room. We talked for a bit. Their optimism was almost palpable.

I retired to my bunk just before 0100, setting my alarm for 0400, our next appointment with AUV 7. When the alarm went, I phoned down to ops and was told they were 'in comms' and that there were 'no issues', so I went back to sleep. The next rendezvous was at 1000 hours and, again, all was well and AUV 7 resumed autonomous silent running.

The following one was at 1300. There were the usual sensor checks and position updates; a glitch was recorded on the forward altimeter sensor, but this was a routine matter which was immediately corrected. By 1340, communications had ceased and AUV 7 was back on task. Expectations within the team were soaring.

A further rendezvous was planned for 1830 hours, by which time AUV 7 should have been on line eight of 11 lines. Like everybody, I was brimming with confidence and any concerns I now had were entirely focused on the recovery, which was going to be challenging. So assured was I of the submersible's progress that, instead of attending the handshake, I went for dinner.

I had just finished eating and was walking down the stairwell to the ops room when I ran into Todd Oxner, coming up two steps at a time to find me. Todd is probably the most placid guy on the ship, and if anything ruffles him he never lets it show, but at that moment he had a face on him like a slapped bum. Even before he opened his mouth I felt a chill run through me.

'It's missed its rendezvous,' was all he said. No more was needed. There is only one flight between the mess deck and the science deck and together we ran down those steps like a couple of pelted hens.

Claire Samuel was already there; her face was drained. Devon, Espen and Blake were also there, standing but bent forward towards the screens and looking stony. Behind them, peering over their shoulders, was the data analyst Pierre Le Gall, glasses up on his forehead and mouth agape in disbelief.

My old friend Chad was seated at the centre, in front of the main screen. He wasn't smiling. Chad's always smiling. He looked up. 'Seven's gone AWOL,' was all he said.

Suddenly the door slammed open with such force that the screens flickered. Channing, the AUV team leader, stormed in. 'Right,' he snapped out, 'from now on I want every detail logged, because if it doesn't turn up in the next few minutes there will be an enquiry.'

It was the beginning of a very long, very tense evening.

We are now locked within the hard crust of the most deranged sea on earth. A world of ice without end. Temperatures are plummeting, the ship is struggling and the pack is tightening. Winter is no longer on its way – it is here. What a time and what a place for our world to implode.

For almost two hours, moon pool open, transponders down, we clung by our fingertips to the rendezvous coordinates over the great watery void beneath our keel – waiting, waiting, waiting – but 7 never came.

During those two hours there was much corkscrew thinking and a quite impassioned exchange of views. Head office in London took the lead and consulted with the manufacturers in Norway and experts in Texas, where the AUV fleet was headquartered. The first hypothesis was that the HiPAP had failed and that 7 had simply slipped by unnoticed. The second involved a range of possibilities, with human fallibility as their common denominator. In this regard our foremost thought was that 7's inertial navigation system might have been given faulty positional instructions, but when these and all the other 'operator error' permutations had been checked and eliminated, attention moved to the vehicle itself. Was there, as somebody put it, a worm within our technological apple? We went through all the pre-dive diagnostics and system checks, but there were no irregularities of any kind.

We needed to know what had incited 7 into going rogue. An AUV has a prescribed set of responses to a range of adverse events, both internal and external, so if we could frame a rational hypothesis for what had occurred then, from that, we could make deductions regarding the vehicle's reaction – which, in turn, would allow us to make some fairly reliable determinations as to its current behavioural mode. And if we could manage that, we would be able to compute its likely present whereabouts.

The greater probability was that 7 was still within the search box so, with that as our baseline assumption, we explored various propositions. Without knowing precisely what had caused it to deviate from its mission, the favoured opinion was that it had gone into backtracking mode. This might happen if, for instance, the FLS (forward-looking sonar) encountered an obstacle. In such circumstances the vehicle would take evasive action, executing an upward vertical loop and then resuming course 10 metres or so higher. If necessary it would perform this movement several times and then, if passage was still impeded, it would most likely backtrack to the loiter box where the mission began.

The second most favoured scenario was that, although we had been on time and were precisely over our coordinates, AUV 7 might have somehow slipped by without making contact. Maybe, for instance, it had experienced a beacon failure. This had occasionally happened on other projects. In that case, 7 would continue forward-tracking until reaching the terminal loiter box in the northwest corner of the search area, where it would have arrived today at about 0700 hours. In our absence it would go into a holding pattern until either we turned up or its batteries expired.

Although the greater likelihood was that AUV 7 was still within the search box it was entirely possible that because of, say, propulsion problems or a failure of the control computer, it might have gone into what is called 'drift' mode. If that was the case it would be at the mercy of the water column and would probably follow the current out of the search box, maintaining altitude until the batteries expired, at which point it would drop its disposable ballast and ascend. If that happened, 7 could now be anywhere under the ice.

Another possibility – and everybody agreed that this was less credible, because nobody could remember it ever happening before – was that 7's propulsion system had experienced a complete failure and it had tobogganed down into the mud. This would mean it would most likely be within the search box and, while it had power, its beacon would be sending out signals that we would receive as long as we were within range.

Finally, for any number of reasons, 7 might have aborted mission, dropped its weights and returned to the surface. This I had seen happen many times before. If that was right, it would be lodged somewhere beneath the floes and possibly still within the search box, or its near vicinity. Although it would be out of communication with the HiPAP, which can only send and receive in a fan-like, downward-looking direction, we did have an over-the-side drop-transponder which would be able to pick up any pings coming horizontally from beneath the ice.

Our favoured scenario was that AUV 7 was backtracking to the loiter station in the southeast corner of the search box where it had started its mission (although its batteries would probably expire shortly before it got there). If, indeed, this was what was happening, then we should be able to perform an interception. We knew its speed and the length of the lines, and we had time brackets for the period within which it would have performed its reverse turn. Channing did the maths and worked out that we had to be over line five.

The problem, however, was reaching the interception point in time. It was just a short distance southwards but we would be breaking ice, charging, reversing, charging every inch of the way. Since entering the pack, everything had become a challenge. We were in hostile territory and facing a number of variables that none of the AUV team had ever had to contend with before – and right now they all seemed to be conspiring against us.

To begin with, there was the manner in which the tide affects the consolidation of the pack, so that it is constantly clutching and unclutching. This was something Shackleton does not seem to have noticed; indeed, even our captain and ice skipper were not previously aware of this phenomenon within the context of the pack. And then there are the consequences of the temperature swings. Between midnight and 0600 there is a dramatic drop that cements the floes and curdles the water within any openings. In other words, if we were on a falling tide at night and the temperature was dropping below $-10°C$, we would be in trouble.

The type of ice we have been encountering within the pack has also been a major hindrance to progress. This ship can take on most ice fields, but not the compacted, thicker, hummocky, older stuff that is commonplace here. This kind of ice is much stiffer than younger ice because of the relative absence of brine within its fabric. We can throw ourselves against this gnarled-up multiyear pack all we like but unless it contains fissures, we will make little headway.

Another factor we monitor closely is drift. Since entering the pack it has been moving between a northeast and a northwest direction at less than a knot, but this can increase to three knots depending on wind and current. And finally there are the unpredictable weather bombs that just come out of nowhere and whack us. Low, rolling fogs, downward-blasting katabatic winds and above all, whiteouts: vast curtains of snow that smother the ship and reduce visibility to zero. Tide, temperature, weather, ice conditions: when they unite into a single antagonistic force, it feels as if we have all four riders of the apocalypse bearing down upon us, swords drawn.

By 2030 hours we were all agreed that AUV 7 would not, as Chad put it, be 'clocking in'. At that point the *Agulhas II* set off in a southerly direction for line five where, as long as 7 was in back-tracking mode, we should be able to perform an interception. Temperatures were beginning their nightly drop and everything around us was hardening.

Progress was slow, but by the second half of the evening we had reached line five and there we waited. Not a single ping rose from the void. We began moving the ship in a westerly direction, but as time passed everybody, one by one, came to the conclusion that 7 was not backtracking and therefore (unless it was in drift mode or had lost motive power) must still be proceeding, as programmed, towards the terminal rendezvous box.

If we did not arrive on time and it ran out of power, it would automatically jettison the solid steel disposable ballast and behave in one of three ways – none of them good. First, it might rise to the surface where, very likely, it would become trapped beneath the sea ice. In such a position the signals from the Iridium satellite position-fixing system would not work (because its waves cannot pass through water) and our chances of finding 7 would be extremely thin.

Second, it might sink to the seabed. AUVs carry both permanent and disposable ballast. Normally, once the disposable ballast has been released to the seabed, the vehicle automatically rises to the surface. However, because of the difficulty we've had in diving the AUVs in the Weddell Sea, it has been necessary to add several kilos to 7's permanent ballast. The line between positive and negative buoyancy is very fine and, because of that extra weight, nobody is now sure how it will react once its disposable weights have been dropped. It might have a slight negative imbalance and sink. In the normal way of things we would send down the ROV to perform an extraction – but of course we no longer have an ROV.

The third alternative, equally bleak, is that following the release of the disposable ballast the AUV remains neutrally buoyant and, in an inert state, simply drifts off on the current, in which case we will never find it.

From all of this it will be abundantly clear that as we pulled away from line five, the only thing that mattered – in fact, was absolutely essential – was that we should reach the terminal loiter box before the AUV's batteries expired and it lost motive power. So another critical issue for us at that moment concerned the amount of life left in the batteries. For a mission time of 42 hours, it had been given sufficient charge for almost 60 hours. However, because something had gone wrong, it was conceivable that there had been an additional drain upon the charge which might have reduced the battery life to 48 hours. The AUV has been programmed to reach the terminal rendezvous box at 0700 tomorrow – meaning that, whatever happens, we have to be over the final loiter box in the northwest corner of the search box by no later than 1300 hours.

It is only a short distance away and we have over half a day to get there; in normal circumstances this would be easy, but these are anything but normal circumstances. As we left line five and headed in a north-westerly direction for the terminus on line 11, we could feel the ice muscling up around us. By midnight, we were stuck fast.

NOON POSITION: 68° 38.838' S, 052° 25.914' W

12 FEBRUARY 2019

We battle the ice in search of AUV 7. In need of a miracle

From opposite ends of the bridge John and I exchanged pained expressions of disbelief. It seemed to be a feature of this expedition that just when you thought things could not get any worse, they did.

By the early hours there was a touch of fog that gave everything a clammy, opalescent quality which matched our mood. Exhaustion and despondency were beginning to bite, so the AUV team and the analysts were stood down to get some sleep. Of that group only Channing was left. His face was drawn and pale from lack of sleep but he was still tapping out messages to colleagues in England, Texas and Norway who were brainstorming the situation with us.

Freddie, the ice captain, was in charge of the ship until Captain Knowledge came back on shift at 0600 hours. Grim-faced beneath his dark glasses, he just stood there, arms folded. The pressure was fierce and he was much concerned. Every now and again from somewhere out in the mist came the sound of rupturing ice. Sometimes it was sharp, like a clap of hands, sometimes more muted like distant thunder. Previously, the clutching of the pack was always followed by the de-clutching, but this felt and sounded different. It was as if we were in the jaws of a vice and some evil ice spirit had given the screw an additional turn.

Whatever happened, we could not allow the ice to establish a permanent grip on our hull. Freddie decided to sally ship once again. The crane operator was summoned and 10 minutes later I watched as he clambered up into his control booth some 15 metres above the deck and energized his machinery. Its wires went taut, everything flexed and a huge wodge of snow tumbled from its jib onto the deck.

The hoist was lowered and a deckhand manipulated the hook to engage with the 16-ton tank of helicopter fuel that straddled the centre line of the well deck. Carefully it was raised from its cradle and then, on an extended boom, was slowly swung in an arc that passed in front of the bow, from one side of the ship to the other. At the same time the vessel's liquid ballast was pumped between bow and stern so that the ship was pitching as well as rolling.

Everything now depended upon the tide and the rise in temperature that came with day. There was a tidal range of 1.8 metres but – and this was a matter of utmost concern – we were coming into neaps, when there was much less movement between high and low water. The horror that nobody dared voice concerned what would happen if, when we reached high water and maximum daytime temperatures, we were still fast. With the approach of winter and temperatures plunging there was, for the first time, a real danger of the ice consolidating into a slab around us and if that happened, we could end up being cryogenically preserved within the pack until spring.

It must have been about 0200 hours when, unexpectedly and to our huge relief, a long crack appeared in the ice before our bows. Freddie went for it. The ship wedged its way forward and elbowed it open. The problem now was the fog, which had reduced visibility to about 100 metres. There was a feature showing on the radar that *might* have been a large opening, but we needed confirmation. Orlando and Vince, the vessel's drone pilots, launched one of their quadcopter UAVs (unmanned aerial vehicles), what they called 'our eyeballs in the sky' – it was soon lost to view, but within minutes was sending back images of a major pool up ahead. This was now our objective, and within an hour we had broken through.

Although we called it a pool, it was really an area where the ice had separated on the rising tide, revealing water that had congealed into a viscous porridge of grease and nilas ice. We traversed this and were soon once more breasting our way through the floes. By now we were over line seven; a wind had arisen, thinning the fog, and our speed was up to two knots, which in this environment was good progress. If we were able to continue at that rate we would make the terminal loiter station by 0700, AUV 7's estimated time of arrival.

But then conditions deteriorated further; the jigsaw of floes around us began to tighten and once again the *Agulhas II* was struggling. Peering into the mist through his binoculars, one of the officers spotted the mouth of a lead that appeared to be heading in a northerly direction. Freddie worked the ship a few points to starboard and we entered the opening. It consisted of two great embankments of old, heavily snowed ice with a corridor of dark water between. For a while everything went well but then, quite perceptibly, the sides of the channel began to close like the two halves of a converging rugby scrum. The ice buckled and broke against our sides, our wake sealed behind us, and once again we were encircled.

It reminded me of a moment when the *Endurance* had also been seized between two vast expanses of mobile pack. One of the diarists described how 'the sliding ice came up against the ship's side with a tremendous impact, grinding and piling up with an accompaniment of groans and cracks which were not a little alarming. At the same time the ship vibrated from stem to stern, and the masts quivered.'

Suddenly everything seemed very threatening. Through the veil directly to port we could just make out the spectre of a rather intimidating iceberg. Distances in these conditions are hard to judge, but it could not have been much more than 150 metres away. Huge, old, mute, somnolent, its face was smooth like a gravestone worn blank by time. Its top and one shoulder were tinged red by the filtered light from the midnight sun. Around its foot was a bulldozed bow-wave of broken sea ice.

Although we could not see it, we knew that beneath the surface this berg had a huge keel that was still being pushed along by the stream. Usually I find bergs sweet to the eye and eternally watchable, but as I looked at this one, it occurred to me that we were down current of it – and while it was on the move, we were not.

In Shackleton's *South* there is an uncomfortable moment when the men realize a 'huge domed berg' is slowly coming down upon the icebound *Endurance*. Over a period of days they observe its approach with growing apprehension. The ruptured ice about its base has rafted up to a height of 60 metres. In his diary, Orde-Lees writes, 'Woe betide us if we come into too close proximity with it.' Shackleton, ever cautious, has all the stores ready to be offloaded at a moment's notice: 'It was easy to imagine what would be the fate of the *Endurance* if she entered that area of disturbance. She would be crushed like an egg-shell . . .' I remembered the power with which a berg had driven into a floe just ahead of our bows several weeks ago. If that million-ton monolith of solid water came crashing down upon us, we, like the *Endurance*, could end up on the wrong side of history.

By 0700 hours, when AUV 7 (if it was forward-tracking) should have been closing on its end-of-mission loiter box, we were still stuck. Our only hope now was the tide, which, later in the morning, would be on the rise. Freddie was now off duty and Captain Knowledge and the mate were in charge; Knowledge, brows knitted, pored over satellite imagery at the chart table while the mate was at the control console.

Soon we began to feel a slight loosening of the pack as the rising tide pushed up on it from below. The mate went over for a few words with the captain, then returned to the console. 'Time to fire up this bitch,' he said, more to himself than to anyone listening. About eight of us, including the captain, moved up behind the glass at the front of the bridge for a better perspective as the ship roused herself to life and began to apply pressure on the ice ahead. Something gave and we stuttered forward 20 or so metres, then ground to a stop. The vessel was then reversed; there was a pause, and then the mate threw the *Agulhas* at the ice as if it was a wrecking ball, a meat grinder and the Hammer of Thor all rolled into one.

Despite the high stakes, I looked on with fascination. It felt as if I was watching a battle of wills between the mate and the ice. It was awesome – and then, at last, we were on our way, wriggling between the floes or, when that was not possible, chopping through them. As long as there were no thick ice fields to impede our way, there was still just enough time to reach the northwest corner of the search box where, with any luck, 7 should be circling.

But then the fog returned. It came down upon us like a thick, all-enveloping blanket, too thick to send up drones. Four sets of binoculars peered into the blur, trying to read the lay of the ice ahead. All we now had to see our way forward was the ship's ice radar. The captain was tense. Two men in hard hats, heavily dressed against the cold, stood at the bow, radios pinned to their chests so they could communicate with the bridge. Peering at them through binoculars, I could make out the ice on their balaclavas beneath their nostrils where the moisture from their breath had frozen into downward-hanging nodules.

The only person not on his feet at that moment was Channing. He was still in the pilot's swivel chair but his head had slumped to his chest and his body had collapsed as if somebody had stolen his bones. He had not slept since the crisis began over 40 hours earlier. He had given the officer of the watch instructions to wake him the moment anything happened or any messages came through.

Despite everything, we reached our rendezvous coordinates almost exactly on our deadline of 1300 hours. Claire immediately opened the moon pool and lowered the HiPAP. The tension we felt was almost as crushing as the ice. If this project was to be saved, then it had to be now. Should 7 respond, we would instruct it to perform a carefully staged ascent into a patch of open water at our stern that we were already blasting clear with prop wash. And then, once we had downloaded the data, we would know whether we had found the *Endurance*.

For over an hour we listened, but nothing arose from the abyssal bottomlands 3,000 metres below. Nobody said anything, but we could all feel things slipping away from us. A terse message was send back to London and Texas that said it all:

Arrived final loiter location at 1300 UTC. Monitoring
conducted. Still no contact with AUV 7.

There was a thimbleful of hope left. Although highly unlikely,
there was a tiny outside chance that AUV 7 had gone to ground
but with its beacon still bleeping. In particular, there was an area
to the west of the box which had not been covered. So we started
to shove and bludgeon our way in that direction. For some reason
we didn't receive any satellite imagery for studying the conditions
of the ice ahead this morning, so the reconnaissance we were using
was over 12 hours old. As we were moving west, Captain Knowledge
came over to Channing, John and me. In his hands he held the
latest satellite intelligence. From the expression on his face I knew
it wasn't good. Claire was summoned from the ops room and
together we studied the printouts. It could not have been worse.

A very large floe, measuring roughly eight by eight nautical
miles, had muscled its way into the west and southwest end of the
search box and was heading in a northerly direction at a stately
1.2 knots. We had been watching it with interest for a couple of
days, but it had always looked as if it would do no more than clip
us. Evidently it had changed course, for it was now covering
precisely the area we wanted to search, and at the pace it was going
it would not clear the area for a couple of days. Furthermore, it
was ice of the worst possible kind: gnarled, mangled, multiyear
stuff that was five metres thick and as tough as teak.

Captain Knowledge passed the satellite scan to Captain Freddie.
'Not good,' was all he said.

We decided to break through to the giant floe and try to nudge
up beside it, to reach behind and below its rim with our transponders. The trouble was, it did not have a well-defined edge – it
had an irregular, rough and crusty outer margin that was difficult
to interpret, and the closer we got to it, the more dangerous it
looked. Besides, we were by then getting blizzard warnings.

By mid-evening, the captain felt we had reached a point beyond
which we would be endangering the vessel. None of us wanted

the ship-hungry fangs of the floe buried in our bottom so, after one final listen, we came about and started working our way east and then sou'-sou'-east. The revised plan was to go all the way back to the very first loiter box and, from there, retrace the AUV's first lines, pausing to listen every two kilometres in case it had been backtracking, run out of battery and gone down in the mud.

Everybody has been trying hard to be positive, but somehow this latest twist, or what the captain calls our encounter with 'the monster floe', seems to have bled the last vestiges of hope from us. When we pass each other on the landings or in the passageways we have nothing to say; there is none of the usual joviality and chat, just a nod and a weak smile. In fact, weak smiles seem to have gone viral throughout the ship.

As I write this we are approaching midnight and – there is no other way to say it – our situation is grim, very grim. We are in a whiteout, surrounded by hull-lacerating ice in the middle of a frozen wasteland, fighting the almighty power of the pack while searching for a two-metre-long tube, 3,000 metres down, that could now be anywhere within the black, cavernous, sealed tomb beneath our keel.

Never, I venture, has maritime archaeology been so challenging. It seems like a very long time since we were all strutting around optimistically, our chests puffed out and red with excitement. In the click of a finger, all that has gone.

On the day they abandoned the *Endurance* one of the diarists wrote: 'Things have taken a terribly serious turn; our worst fears are realized.' At that moment, all hope of ever achieving their expedition's objectives lay in ruins. And so it is with us; unless there is a miracle within the next few hours, we too have hit the buffers.

NOON POSITION: 68° 36.148' S, 052° 40.679' W

13 FEBRUARY 2019

'It's over!'

Approaching midnight, while trying to make it through the ice from the northern to the southern end of the search box, we again became fast, this time for over four hours. But even when we were free progress was punishingly slow. In just under six hours we covered a mere three kilometres. We all felt the frustration, but nobody voiced it better than the mate: 'I have known barnacles that can move faster than us,' he fumed.

By lunch we had reached the southeast corner of the search box, where AUV 7 had begun its mission. Straight away we lowered both the HiPAP and the over-the-side transponder, but not a peep arose from the pit below.

Our plan was now to follow the route taken by 7 after it had been released from the first loiter box, the idea being that, if it had gone into backtracking mode, then we should come across it in the mud at the point where its batteries had expired. But then, once again, Fate intervened and delivered yet another kicking. It was something that should have had us slapping our foreheads in disbelief, but the events of the last day and a half have left us emotionally numb.

Once more it was the monster floe that thwarted us. Frazer Christie came onto the bridge clutching a small computer that featured the latest satellite intelligence. This showed that during the preceding 12 hours it had altered course and rotated slightly so that it now formed an impenetrable slab over the very sector we wished to search next. Currently it shields more than a third of the search box and, given its heading and speed (which has dropped to less than a knot), it will soon be blotting out most of the box. The only area still open for investigation was a triangular zone along the southern side of the box, but the consistently low and still falling temperatures meant that everything was now coming together and freezing solid. According to Frazer's imagery there were no leads that would allow us to hopscotch our way around within the triangle and, where there were small open pools, the perishing cold had turned the water into grease and nilas ice, giving it the consistency of molasses, which would seriously impact our ship's thrusters.

We battled on for a few more hours but seemed to get nowhere. For the present the skies over the ship were free of mist; they appeared vast and malevolent, and although the sun was high it bestowed no warmth. In whatever direction one looked, the white wasteland about us, though puddled in places, seemed alien and unrelentingly belligerent. Even if we did come across 7 it would most likely be on the seabed and we had no means of executing a recovery.

As we passed high water during the second half of the afternoon we could, again, feel the pressure building and soon, once more, we were clamped. It felt as if the Weddell Sea had both its thumbs on our windpipe and was slowly squeezing the life from us. In the past, when stuck, we've known we would very likely be able to break free as the water rose, but now that we are into neap tides things have become perceptibly more treacherous. The question hanging over us concerns whether we are experiencing one of this region's notorious 'flash freezes', or whether this is part of the ratchetting process that takes us into what might be called the Big Spread: the most prodigiously transformative event on earth.

Sea ice comes and goes with the polar seasons. We are at the back end of the austral summer, when the South Pole leans away from the sun, the long night descends and temperatures drop so low that even the mercury within a thermometer freezes. Soon the sea ice will be building mass across the surface of the sea at a rate of 22 square miles a minute. This will continue until the whole of the Weddell Sea has become a grand massif of stone-cold solid, wall-to-wall water. But it goes beyond that; it will continue to expand until it engirds the continent and forms, as it were, a frozen lid over the entire bottom of the globe: all in all, 7.5 million square miles of congealed ocean, an area twice the size of the United States. If we misjudge things and do not get out of here in time, we will be in serious trouble.

By late afternoon the ship was free, but we were struggling to cope. In an hour and a half we made only a single kilometre. Before we knew it we were approaching evening, when temperatures would plummet and the water would further harden about us. A low mist had returned, wreathing the sun and the trailing clouds in glowing tones from the more ruddy end of the spectrum. There was nothing left to say; we were all drained, we were all out of ideas and, although nobody said it, out of hope. There was a clear sense that we were on the edge.

On the bridge, Captain Knowledge beckoned to Freddie, and for a while the two conferred privately over on the starboard wing. The rest of us knew what it was about. There was a decision to be made, and it was stark. Should we pursue the search for another 24 hours, or should we leave now? Our chances of finding AUV 7 were negligible and Knowledge's first thought had to be for the ship. I watched their mouths move but couldn't hear what they were saying.

As we waited, I looked out over the ice. On the edge of a floe, a little to the south, a penguin stood alone. I remembered reading that when they abandoned the *Endurance*, some emperor penguins waddled over, raised their beaks and 'chanted a weird dirge'.

Then the captain came over to deliver his decision.

And so it was that everything came to an end. It wasn't quite as dramatic as I predicted yesterday. Nobody remonstrated; in fact, nobody said anything. We all just stood there, arms akimbo, alone with our thoughts, either studying the deck or gazing out over the ice into the white distance. Simply put, we had run out of road.

There was a similar moment in the battle to save the *Endurance* when everybody knew they could do no more. For two days they had manned the pumps but were unable to beat back the rising water within the ship. All night they had been pumping and were utterly exhausted. Macklin wrote in his diary: 'I don't remember anyone telling us to stop, but we just seemed to understand that we might as well give up.'

The *Agulhas* was brought about and, under an implausibly streaky sky, we began our retreat. 'Time to get out of here,' said the mate.

For a few minutes, away from the others, John and I talked through all that had to be done next. He now had the unenviable task of writing to the project's management committee and the trustees of the Flotilla Foundation. I then went to my cabin as I also had some difficult calls to make, starting with head office.

On the way I passed through the lounge. All the ROV guys were there, plus some of the AUV team, both data processors, a couple of scientists and several students. I paused, feeling I ought to say something. They all looked towards me and, doleful-eyed like an executioner about to raise his axe, I looked back. This was not the punch line of which I had dreamed. The words seemed to stick in my throat, but I got behind them, gave a push and out they came.

'Ladies and gentlemen,' I said, 'it's over.'

Back in his cabin, John pondered what to tell the team. At 2030 hours he tapped out the following:

Tonight we have decided to stop the search for AUV 7
and exit the *Endurance* search area. This is because the
temperatures have dropped significantly in the last 24 hours
and the open water leads we were working in were all
freezing over, making it very difficult for the ship to break
through or manoeuvre. There was a very real risk of the
ship being nipped between the big floes and becoming
beset . . . we now need to think about the safety of the ship
[as] the Antarctic winter is coming on fast . . .

Had we found the *Endurance* this would have been a day of celebration, but instead it became one of crumbling dreams, pounding
frustration, and introspection. Instead of hallelujahs people will be
questioning our motives (as indeed they did Shackleton's following
his return). From where the idea was first hatched, in a tiny coffee
bar in South Kensington, all the way to here, the hard gut of the
Weddell Sea: why? By any measure this is an absolute hell-hole of
a place, the most hostile spot on earth into which any ship can
deliberately swing its helm. So why did we do it?

Unlike Everest, we were not looking for the *Endurance* because
it was there; nor, like the moon landing, did we do it to become
'the first'. Our purpose was to locate, study, record, interpret, share
and protect. In other words, we were seekers of truth on the shiny
path to knowledge.

And so we have failed. I have failed. Everybody else on this
mission has done what was asked of them; I am the only one who
did not deliver. I cannot pretend it doesn't hurt. But as I mull it
over, another thought arises. Our efforts have not been entirely in
vain. There have been several attempts to mount expeditions to
find the *Endurance*, some by eminent, wealthy organizations armed
with the best know-how and technology; but none of them achieved
lift-off. We have demonstrated that finding the wreck is now technologically feasible. We have thrown open the doors. We have
blazed the trail. Like Scott and Shackleton, we may not have claimed
the prize, but we took those crucial first steps that will enable others
to succeed.

And so to my bunk. History, today, has not been kind to us.
Now I just want to put these last 48 hours behind me, wake afresh
and start again.

NOON POSITION: 68° 43.251' S, 052° 19.460' W

14 FEBRUARY 2019

Despondency. The team goes out on the ice

I was up early as I couldn't sleep. The sky was a camel colour, as if streaked with mustard, the sun low and splodgy. The horizon was much distorted by refraction. Everything felt a bit surreal.

I went up on the bridge and was surprised to find the ship was heading in what I thought was the wrong direction. It took no more than a glance at the satellite imagery to understand what was happening. They were trying to punch through to a series of long, ill-defined leads that were trailing off in the rough direction of the ice edge. If we could connect with them it would shorten our passage out of the pack by at least 10 hours.

As can be imagined, our spirits at breakfast were low. After breakfast the entire ship convened in the auditorium to listen to the captain. John and I were sitting in the front row. Captain Knowledge came in and sat down beside us. John then got up and said a few words of appreciation for a man who had been magnificent in his management of the ship throughout this difficult mission.

The captain himself then rose and, crystal clear and polite as ever, explained the situation and why we had been obliged to resign and withdraw. None of us doubted his decision. I thought of Shackleton on the *Nimrod* expedition, when he got to within a hundred miles of his dream, the South Pole, before having to retreat. That was probably the most difficult decision of his life: there is no doubt that he could have made his objective, but equally, had he pushed on, there is no doubt that at least some of his party (like Scott three years later) would have perished on the long haul back.

After we left the auditorium, the AUV team together with the analysts and myself went down to the ops room to evaluate the events of the last two days and look at what had gone wrong. While we waited for everybody to assemble, AUV team leader Channing sat to one side quietly sipping tea and leafing through his notes. At last he put down his cup with a click that, like a gavel, brought everybody to order. He stood up.

'Everybody here? Right, close the door because I do not want to be disturbed.' He then turned to the whiteboard and in large capitals wrote: 'LESSONS LEARNED'.

For almost two hours, every move and procedure was analysed in minute detail.

The most interesting part of the proceedings concerned what we will do differently if we resume the search in the future. Nobody has attempted under-ice AUV operations in such challenging circumstances before, so we have learned a lot. If ever we do return, our search techniques and operational methodologies will be completely different.

When we left the room it was late morning and we all felt somewhat wrung out but, at the same time, more than a bit relieved. There's no consensus on what exactly went wrong with AUV 7 but we're all in agreement that whatever it was, it was a miscarriage of technology and, simply put, we have been caught by the swing of chance. So you see, m'lud, it wasn't us. Nobody goofed. Sometimes bad things just happen.

That is not to say we haven't made mistakes. We have. There were some software glitches that could have been avoided; a poorly positioned aerial meant that we had a blank patch on our GNSS (global navigation satellite system) for a while; there were ballasting issues; the transceivers on the HiPAP head required modification; certain tools and items of equipment didn't arrive at Cape Town in time; and so on. Our most obvious misstep was allowing a squeeze to arise that reduced the time available for the wet testing of our submersibles. If there was a pattern to our errors, then it was that they almost all occurred during the mobilization phase in South Africa before *Agulhas II* departed for Antarctica.

Scott used to say that 'the worst part of an expedition was over when the preparation was finished'. This expedition has proved he wasn't far wrong. There was much pressure on everybody during this phase and people became stressed.

Following the debriefing I went to the lounge for a coffee. By then it was late morning and we had broken through to the leads. Nobody was talking much, most were just buried in work, and the mood was still leaden. Somehow we had to snap out of it. Someone came up with the idea that as soon as we came upon a stable floe, we would pause the ship and let everybody get down among the penguins and ruffle a few feathers. The scientists could also take cores and dig ice holes to analyse the salinity, crystalline structure and snow content.

Not long after lunch, we came upon a large floe that looked staunch and was flecked here and there with penguins and a couple of seals. We drove our bows up and over its apron and John was lowered by the crane to inspect the crust. I watched him stomp around, probing the upper layers with a stick as he went, a lone red figure in a vast whitewashed desert. The wind had gone down and an unclouded sun sent little prisms of light skipping across the ice towards the ship in a way that made us squint. I thought if ever there was a Lawrence of Arabia of the snows, it was John.

At last he proclaimed it safe, so we all layered up and the crew started winching us over the side and onto the ice in groups of four or five. To be honest, I almost didn't go. The grim necessity of being nice to penguins was too much for me at that moment. In the end, I only went to show willing.

However, as we stepped off the foot-ring and felt the crunch of snow underfoot, something rather unexpected happened. An insufferably precocious little brushtail, on its way to the water, stopped a few feet in front of us, raised its bum in our direction, defecated in a single long splurt, then toddled off. We laughed. The mood was broken. Then we laughed some more. At that moment, I knew we had all put the disappointment of the last couple of days behind us.

The scientists began their testing and another cricket game started up. Then an emperor penguin was suddenly among us: head high, shiny wet and full-chested with food. As if he had all the time in the world, he sauntered about between us with a certain self-consciously affected nonchalance that reminded me of a Parisian boulevardier from the early last century. In the end everybody had to stop what they were doing; the iceniks got out of their holes, the cricket game stopped and we all lined up in an irregular half circle around him, cameras clicking while he obligingly came right up and inspected us all, one at a time.

After half an hour or so, Captain Knowledge and the ice skipper, Freddie, were beginning to get twitchy. They had just heard that a major storm system was heading our way and they wanted to be out of the pack before it struck, so it was all back on board.

Before turning in tonight, I wandered down to the stern for a moment alone. I was thinking about how we'd come to Antarctica to expose and explain one of its greatest enigmas but instead, all we had done was add to it yet another layer of mystification. I had not been down to this part of the ship since we launched AUV 7 and was momentarily startled by the empty cradle before me, where it should have been.

I've spent a lot of time with 7 on some great deep-ocean adventures in the Indian and Atlantic Oceans. I know it's just a robot, but you become attached to them, as you do to a car; and so I feel a little mournful over its loss. At the same time, I have no doubt that one day it will be found. It may wash up, barnacle-encrusted, on some distant shore, but if not, at the rate deep-ocean sensing technology is advancing, it will be discovered by know-how and applied science that is currently beyond my imagination.

Behind its nose cone, there is a little plaque engraved with contact details and the promise of a reward for its return. I feel certain that this will not occur within my lifetime but when eventually it happens, I fervently hope that AUV 7 ends up in a museum as a tribute to this expedition and the science of our era.

NOON POSITION: 69° 02.619' S, 050° 09.668' W

15 FEBRUARY 2019

Shackleton's birthday. Telling the world

Just when we were starting to think we might have been given a free pass out of here, the ship is once again wrestling with our frozen environment.

Overnight, ice conditions deteriorated sharply. I first became aware of the situation yesterday evening, when we hove to with our nose into the wind to take seabed sediment cores. Annoyingly, the winch malfunctioned. This was a fairly routine matter but when the bridge was informed that we would require an additional 45 minutes to make good our defects, their prompt response was to stand down all coring operations as the ship was resuming passage 'with immediate effect'. No explanation was given, but I knew if they could not spare so little time then something not good was stirring. And it didn't take much to deduce that it was the ice.

I went up on the bridge. There was no chat. Everybody was at their posts and studiously attending to their duties, but you could feel the tension. Since I was last there the panorama had changed completely; the ice-spreads around us were now dominated by thick, gnarled, multiyear pack. There isn't a ship on this earth that could head-butt its way through that stuff. From the radar, however, I could see that we were not within a solid frozen crust, but rather at the centre of a treacherous-looking labyrinth of pools and minor alleyways – and that connecting these openings to get out of here was going to be a challenge. Soon the ship was back and forth all over the place looking for outlets. It shoved and shunted, it dodged and weaved, it zigged and it zagged, charging here, breaking there; it was relentless. The unspoken maxim was stay mobile, stay nimble, keep walloping and, whatever happens, don't stop.

Soon after, John put out a message to the team that said it all:

The ship is battling through very tough ice conditions. She
will have to back and ram constantly. Captain Freddie has
decided that we must exit the pack right now to ensure the
safety of the ship. All planned [science] activities are
postponed until we have reached the ice edge. We should
get there at about midnight tonight.

And so it went, hour after bloody hour, what a slog; but in the
end, thanks to the icecraft of the two skippers and the mate, we
got through. In the saloon you could feel the relief.

It was Shackleton's birthday today and head office in London gave
permission to open the bar after dinner. As usual, we were each
allowed two beers or glasses of wine. I spoke for a few minutes on
the great man's legacy and then handed over to John, who, with
his usual élan, proposed a toast 'to the sacred memory'.

How did Shackleton's men celebrate his birthday?

His first birthday in the Weddell Sea (when he turned 42 years
old) was a particularly backbreaking affair. At the time the *Endurance*
was jam-packed solid within the ice but then, in 'Jock' Wordie's
words: 'Some pools appeared a little distance ahead . . . our object
is now to [use them] to reach a bad lead, which is probably 300 yds
distance away; this lead ultimately becomes a good lead which runs
as far as the horizon.' On the day before Shackleton's birthday, they
began the task of trying to cut their way through to the first of
the pools.

They worked throughout the evening and then next morning
they recommenced work, cutting up the ice into triangles which
were then poled to the stern to fill the tract left by the vessel's
forward progress. An hour before midday, the ship broke through
'the ****** Lump' and progressed into the artificial pool they had

created. By then they were beginning to doubt the wisdom of their labours.

Shackleton makes no mention in *South* that this was his birthday but, interestingly, he did think it a pivotal moment in the *Endurance* story. Although he had not given up all hope, this was the day upon which his dream of crossing Antarctica came to an end and, in his words, he began 'counting on the possibility of having to spend a winter in the inhospitable arms of the pack'.

Shackleton's second birthday in the Weddell Sea was a miserable affair, the highlight of the day being the capture of an emperor penguin. By then, the men had been living on the ice for over four months and were worn down and weary. They had been looking forward to the occasion as an excuse to splurge a little. Orde-Lees wrote that they 'had arranged a spread magnificent in proportion to the few stores we have left, but on broaching the subject to him he most emphatically declined to allow the anniversary to be marked in any way, although he appreciated our intentions'.

The following day, Orde-Lees wrote:

A disgustingly wet day and Sir Ernest's birthday. The strong northerly wind burst the flimsy eight-man pole tent. The driving sleet came in everywhere. Greenstreet and Clark patched up the rent, one inside and the other outside, passing the needle from one to the other through the tent wall. No sooner had they repaired it than the bally thing burst at quite another place and they had to go through the whole laborious process all over again . . . Although Sir Ernest put the kybosh on the prospective banquet, we had the long-anticipated dog-food bannocks for luncheon and a howling success they were. Sir Ernest has quite seriously protested against the size of these bannocks, objecting that they were uneconomical.

Bannocks, it should be explained, were made from flour, fat, water, salt and baking powder. The resulting dough was rolled out into flat patties and then baked for about 10 minutes on a hot sheet of iron over the fire.

Today, Shackleton's birthday, we had hoped to spring a great discovery on the world, but instead the media were informed by press release of our defeat. I doubt they will be pleased. I promised fireworks but have delivered only damp squibs.

I suppose what it really comes down to is that I do not want to be remembered as the person who led this incredible search down what has turned out to be one of history's great cul-de-sacs. Sometimes, in archaeology, it is easy to bury your mistakes and failures – but not when they have been trumpeted across all the papers.

NOON POSITION: 69° 03.853' S, 048° 48.065' W

16 FEBRUARY 2019

Out of the pack

By yesterday evening we were out of the danger zone and into the stretch, a featureless apron of young ice that continues until open water. Aware that I will probably never again experience the Weddell Sea pack, I wanted to relish these last moments. I was on the side deck, the thermometer was tipping $-15°C$, and even though I was wearing gloves, my fingers were starting to ache: a clear message that it was time to get back in the warmth or risk frostbite.

Ray appeared at the watertight door beside me. He was looking a little bleary-eyed, as if he had just woken up. He fumbled for a cigarette. 'Hey, bub,' he greeted me, and then, looking around, added: 'Where the hell are we?'

I smiled and gave him his cue. 'No idea. Must be lost.'

He put on an exaggerated Brooklyn accent and imitated Bugs Bunny eating a carrot: 'I told you we shoulda taken that left toin at Albukoykee.' We both laughed.

It is difficult to be precise about when we actually broke out of the pack because it did not have a defined edge, it just sort of crumbled into what Shackleton's men called 'bergy bits'. At one stage I thought we were clear, as the sea, which had been flat within the openings of the pack, was once more furrowed. Within half an hour, however, we were back into a large orbital field of snaggle-toothed ice that once again had us breaking. However, just as quickly as we were into that we were out of it, and as we head for Penguin Bukta we have raised revolutions until we are cantering along at a dazzling 14 knots.

And so, to quote the mate – who has a euphemism for everything – we are now 'out of the barrel'. I should feel relieved.

And I do, but I am also a touch wistful. I can't quite explain it, but I should be a bit more cheerful than I am. When we were locked within the pack, or within what Shackleton called 'the worst portion of the worst sea in the world', we were emotionally wrung out, hung out, pulled end from end and pummelled. At times it was as if a band of little ice demons were gleefully sandpapering my nerve endings.

But – and this, in hindsight, is what surprises me – I have never felt so alive in my life. The whole thing was utterly intoxicating.

NOON POSITION: 68° 50.111' S, 042° 12.615' W

17 FEBRUARY 2019

In medio tempestatum

And then the storm came.

Of course, we knew it was coming. A couple of days ago we were told that a major low-pressure front was on its way. By yesterday, the isobars were packing in, the barometer was plummeting and it was obvious that we were on the cusp of something big. Rule of thumb: the deeper the depression, the stronger the wind and the worse the storm. 'Going to be a nasty bastard,' said the mate.

Alerts were issued but I never gave them much thought until yesterday afternoon, when we received a rather breezy warning on the team messaging service from chief scientist Sarah Fawcett:

Hi everyone. The bridge has just let me know that we are about to hit some bad weather . . . please secure all scientific kit in labs and on deck.

Soon after that, the ship's meteorologist was running around in a flap telling us to ensure our computers and cameras were firmly fixed or stowed.

Today I got up early and went out on the helicopter deck, from where you have an uninterrupted view in every direction except forward. It seemed eerily quiet. The sky was clear, the sea was low, and although the sun was out, it bestowed no warmth. There was menace in the air. Even the Adélies that were gathered in small, subdued conclaves on floes seemed to feel it. It was as if everything was holding its breath.

Suddenly a puff of wind lifted my hair and I knew that whatever was brewing would soon be upon us. Would it, I wondered, be slow to build, or would it be up and at us like the proverbial bat out of hell?

We pressed east. Before the hour was out the wind was spooling in from the north, the waves were massing and the skies had turned gunmetal grey. In the distant offing to windward, a grim bank of mackerel-blue cloud was barrelling towards us. After breakfast John put out a message to the team:

> Weather conditions getting steadily worse. Wind speed currently 35 knots and predicted to 45 knots this afternoon and then 50 knots. Worse tomorrow.

It used to be said that the sea was a ruthlessly hostile environment upon which you only ventured at your peril. These days, it is no longer the dangerous place it once was. During my late teens, when I was a seaman in the Roaring Forties and Fearsome Fifties of the South Atlantic (these, by the way, are latitudes, not decades), you could not predict the storms with any confidence until they were almost upon you – and then all you could do was buckle down and take the biffing. Nowadays, by contrast, we practise what is called 'weather routing'. Thanks to satellites, storm forecasting has become so dependable that you can predict their arrival and scale with such precision that ships can steer clear of their centres or, indeed, avoid them altogether.

But not here, today, in the Weddell Sea. On our constantly updating electronic weather charts, which give and predict sea conditions all over the globe, the most violent storm areas are coloured turquoise. Today the entire Weddell Sea is lit up in turquoise. Within this vast oceanic basin there are no havens where we can find shelter. Behind us is the pack; the waters to the south are completely ice-clogged; and if we head north, we will be into the 'Screaming Sixties' south of Cape Horn, the most consistently savage seas on earth. Here the winds and currents circulate west to east around the entire Antarctica continent, giving waves thousands of miles of uninterrupted fetch and long runs in which they can overtake each other and conflate into great rollers. The safest thing that we on the *Agulhas* can do is stay within this patch of relatively open water and, as they say in the Falklands, 'dig deep and hold on tight like a Shag Island Harbour sheep tick at shearing'.

By afternoon, waves streaked with foam were lashing at our sides and lifting our bow high out of the sea before letting it slam back down in a pile-driving plunge, sending up great scoops of water that deluged the ship all the way up to the bridge. Out on deck, wind howled through the stays and sleet mixed with spray gave the air an almost liquid consistency. You could even taste the salt. At one point I was joined by Julien Trincali, a keen yachtsman. 'If this were my boat,' he said, 'we would now be flying along at over 30 knots under a single staysail.'

It is as if the Weddell Sea wants to give us one last kicking. We presumed upon its secrets, and for that we must pay. It was not enough that it crushed our project; apparently it intends to horsewhip us all the way back to Penguin Bukta.

By dinner, the pitch and roll had become more rhythmic. That is not to say the storm had eased, but rather that it was no longer deteriorating. Which was just as well, because this was the evening that had been scheduled for the *Agulhas II* Film Festival.

There is a long tradition of amateur theatricals on polar expeditions. During Scott's *Discovery* expedition of 1902, they put on a midwinter musical revue and a one-act 'screaming comedy' in their hut at McMurdo Sound. Shackleton's men, icebound on the *Endurance* 13 years later, marked the day with a show in which everybody (apart from the unfairly excluded crew) performed a sketch or a musical number. For the occasion Hurley built a stage which was illuminated with acetylene footlights. The programme began with Shackleton, who gave 'an egoistic and satiric harangue which [was] admirably responded to by the Rev. Dr Bubblinglove [Orde-Lees]', who extolled the joys and rewards of virtuous living. The jokes were topical, everybody took a ribbing, nothing was off limits, and there were, of course, the usual sexual insinuations, always in the worst possible taste, coming from men in petticoats. Engineer Lewis Rickinson was a flapper, Hubert Hudson was a coquettish captain's daughter, and Dr James McIlroy minced about as an extravagantly made-up Belle Époque streetwalker.

So it was not out of place, and even to be expected, that on this polar expedition we should hold some amateur theatricals of our own. In keeping with our times it was decided by the organizers that we should divide up into small groups, each of which would write and film a brief sketch that would be screened and voted upon at a special gala event to be held in the auditorium on the return voyage.

Tonight was the night. Despite the constant swaying within our seats and the queasiness that some were feeling, it was an infectiously jolly affair.

I am back in my quarters now, finding it impossible to sleep. Around dinner I had the feeling that the storm had plateaued but now, nearing midnight, it has found fresh reserves and is once more escalating. I had another look at my copy of South, curious to see what Shackleton says about the tempest that rocked their crossing from Elephant Island to South Georgia. And so, as we are carried off into the night like so much driven chaff, I leave you with this from the man who said he wasn't a poet:

Deep seemed the valleys when we lay between the reeling seas. High were the hills when we perched momentarily on the tops of giant combers.

NOON POSITION: 69° 02.351' S, 029° 12.320' W

18 FEBRUARY 2019

Freak waves

Throughout the night, conditions continued to deteriorate, and now our ship is in pain. From my starboard porthole I watched as gloomy hillsides of water rose and fell about us, swatting us back and forth as if we were no more than a feathered shuttle in a game of badminton. In my bunk I attempted to limit my roll by jamming bags, cushions and life jackets between myself and the sideboards, but still it felt as if there were forces at work that wanted to tip me onto the floor.

And then there were all the noises. There was constant knocking, and every time the vessel flexed there was a chorus of creaks from the panelling against a backwash of more hard-to-place murmurs and grunts that seemed to arise as if from beneath some distant eiderdown. And finally there were the sounds of my own making. I kind of knew what these were, but couldn't summon the animus to get up and do anything about them – the pen on the desk, the bottle in the bathroom cabinet, the thing in the drawer. Above all, there was the apple I took from the galley yesterday, which had fallen from the shelf and was rolling about the floor of my day room.

There were quite a few empty seats at breakfast. Some of the absentees I afterwards found in the lounge, all with the kind of stony expression one associates with seasickness. No sooner had I sat down than my phone bleeped. It was John messaging the team:

The ship is now in a full gale-force storm . . . consequently
we have had to slow on our casterly track – now making
6 knots . . . It is dangerous to go outside with such big seas

running and gale-force winds battering the ship. Keep
watertight doors shut. No one to go out on back deck
without permission from bridge.

Not long after that, there was a message from the captain
ordering everybody to stay inside. 'It is dangerous and the weather
is not improving,' he concluded.

Conditions had indeed much deteriorated and we were in the
grip of a force 10 building to an 11, or what we used to call when
I was at sea in the '60s 'a right rip-snorter'. Wind speeds were way
up in the high fifties, gusting deep into the sixties. Waves the size
of houses were cresting in long cats' paws from which the foam
was scooped up by the wind and sent flying. We were nose into
the weather and when the big ones buried our bows the ocean
came cascading down onto the well deck and snarled about,
swamping everything in spray and white water. And with the gale
came sleet – slashing sheets of it that reduced visibility to almost
nil. But there were also breathers, when the driven white sludge
cleared and we watched in awe as the storm tossed us about between
its folds and interfolds as if we were no more than nothing.

The sea is a lonely place, but never more so than in a storm.
As long as you are on or near the sea-lanes, there is usually another
ship somewhere over the horizon that, if necessary, will come to
your aid; but not here in the Weddell Sea, and certainly not at this
time of year. Until a few days ago there were a couple of ice-
reinforced supply ships – the pit ponies of Antarctica – at work
relieving the scientific stations along the largely open-water areas
of the eastern Weddell Main, but with winter upon us and the
great freeze setting in, these have gone. We are utterly alone.

I heard somebody say that the closest human beings to us right
now are those on board the International Space Station. I suspect
there is an element of exaggeration to that but nonetheless, it sums
up our sense of isolation rather well.

Even so, I am not worried. Not in this ship, which was designed
to absorb such punishment. These days the only thing I fear is the

100-foot, Hokusai-style wave which, although I have never seen it, I have experienced. It used to be that people thought freak waves, or what are more properly called 'extreme' waves, were no more than an overly vivid bit of nautical folklore, but now we know them to be very real and surprisingly common. The record for an extreme wave, from trough to crest, is a 95-footer (29 metres) recorded by a research ship in February 2000 near Rockall. Waves often come in trains. According to data gathered by the British National Institute of Oceanography, one in every 23 waves will be twice the size of the average wave in the train, while – and this is the alarming bit – one in 300,000 will be four times greater than the average wave. In a storm situation, this would be the legendary 100-footer.

The good news is that 300,000 waves is a hell of a lot of waves, and the chances of you being there to see one are tiny. But it does happen, as I found out in February 2015, when I was leading a deep-ocean search-and-survey mission in an old, leaky trawler that had once been a Cold War submarine chaser.

At the time we were between the Falklands and South Georgia, not far to the north of where we are now, and at the centre of a gale-force storm. I was working at the chart table behind the bridge and so never saw it coming. I only realized that my life expectancy had plummeted to virtually zero when I heard the panic-stricken mate yell 'Hold on!' I caught a glimpse of the eye-popping terror on his face and grabbed the stanchion beside me. 'Wall of water!' he bawled. Even as he said it, I could feel the vessel being lifted on its rise. An instant later we were slammed sideways and driven over on our beam ends. The violence was such that navigation and communication devices which had been secured to the fore and aft bulkheads were ripped from their fastenings and sent flying across the ship from port to starboard. I couldn't hold on, and also went hurtling across the bridge. For what seemed like an eternity but could only have been seconds, the ship just hung there while the mate, a seaman and myself lay in crumpled heaps among the debris of radios, printers, computers and weather faxes, waiting for the follow-up wave that would roll us under. We felt the ship fall away into the hole behind the first body of water, then rise again upon the lee of the next. The seaman from Honduras was praying in Spanish. We knew what would happen. It would come down on us like a fist. Over we would go. Down we would go.

But it never crested. We rode its summit and, very slowly, the vessel righted itself. One engine had gone, but the other managed to give us a little bite at the stern and gradually we were able to bring her nose around and into the storm. It was one of the narrowest escapes of my life.

Fifteen years or so earlier, an old trawler called the *Sudur Havid* had been swallowed down in a storm not far from where we were hit by the extreme wave, but closer to South Georgia. Seventeen people died in that incident; I was in Port Stanley when they brought in the survivors. It was a horrifying story that was later told in a book called *Last Man Off*.

Our current storm, however, though it is fierce, is not a big-ship-eater. The difference between this place and anywhere else in a storm is that the Weddell Sea has fangs, or rather great chunks of *Titanic*-disembowelling ice. So I was not surprised when I went up on the bridge a little while ago and found them all with binoculars drawn, scanning the waters ahead for growlers, those 20-ton slabs that have no radar presence but might be tucked away in the dips and could, without warning, rise up and smack us in the gob like a hockey puck.

There was no sleet, but the skies were ash-grey and visibility, at best, was moderate. The mood was taut and there was no unnecessary talk. Everyone just stood there, legs slightly apart, straining to spy any potential hazards within the unlovely billows ahead. They made me think of the monks on Skellig Michael, scanning the horizon for Vikings.

NOON POSITION: 69° 07.828' S, 021° 15.413' W

19 FEBRUARY 2019

The storm expires. The ice passage through to Penguin Bukta

'It was like watching a slow puncture,' said Captain Freddie, who had been on the bridge overnight as the storm wound down.

The time was 0700; the winds had expired, a heavy swell was running and the skies, though still bruised and frowning, were returning to normality. There were patches where the sea was knuckled with ice, but nothing that would threaten the ship, which was now pounding along at maximum speed to make up for lost time. 'If we can maintain this pace we should be into the ice passage by evening, and back at Penguin Bukta sometime tomorrow,' declared Freddie as he passed me the latest satellite images. They showed the shoulder of the Weddell Sea dense with bergs.

Back on the bridge after breakfast, my phone bleeped. It was John transmitting the bosun's list of so-called 'impairments' to the ship:

The big waves damaged the ship with both mast and stern
lights broken and also one of the AUV stinger chains . . .

It was the last bit that grabbed my attention. The stinger is the huge, steel-framed launch and recovery mechanism at the very back of the ship. To have damaged it like that would have required a mighty act of nature. Because the aft deck was sealed off at the height of the storm, nobody saw what happened, but clearly it must have occurred when a massive sea curled up over the stern and then came cudgelling down upon it. It was probably a 'transient' wave. These occur when the wind changes direction and sets in motion a series of contending wave trains. Where two or more of these trains converge, a single wave can mountain up and, as its gradient exceeds, say, 20°, it crests, releasing all its contained energy in tons of crashing water. If the rear of the vessel happened to be in the way at that moment – or worse, if it was in the process of being buoyed upwards on equally prodigious forces from below – then our stinger would have been caught in the middle, as if between an anvil and its descending hammer.

By lunch everything was once again perfect. The sea was flat and sparkling and, apart from a few low cream puffs with some wisps of cirrus in the distant blue above, the sky was open and radiant. The mood at lunch was bouncy. We have the smell of home in our nostrils. For 50 days we have lived the life Antarctic, but now it is almost over. I shall be sorry to leave this ship. The sea has always been my calling and were it not for my family I would happily sign on to the *Flying Dutchman*, the legendary ghost ship that is doomed never to make port or see land.

In late afternoon I was in my quarters sitting at my desk, deep in thought and staring at the sky through my side-port, when all of a sudden, in the words of the poet, 'ice, mast-high came floating by'.

And so our transit of the open Weddell Sea has reached its end. We are back in the berg-haunted waters that mark the ice passage through to Penguin Bukta, where, an age ago, our odyssey began. Penguin Bukta, it will be remembered, is that part of the Fimbul ice shelf where the South Africans on- and offload personnel and supplies for their Queen Maud Land scientific base, Sanae IV, some 80 miles inland. Once we have arrived, the team will disembark in relays that will be flown back to the airstrip at Wolf's Fang for connecting flights by Lear jet to Cape Town. Ideally, we would like to put the ship's nose right up against the glacier and then winch everybody up onto its lip; but if we cannot get all the way through to the cliff face, we will use the low-lying fast ice as a quayside and helicopter the team onto the shelf.

We talk of the 'ice passage' as if it were some kind of strait or oceanic corridor, but it is nothing of the kind. There is no defined route, you just make it up as you go along, threading your way between the frozen formations with increasing vigilance because the closer you draw to Penguin Bukta, the more congested it becomes. Whatever happens, with the 'big freeze' on its way, we must not allow ourselves to become enmeshed.

Sure enough, as the ice began to thicken and meld, our course became increasingly irregular again and fraught with challenge. There were great bergs and floes all around. It was as if we were in a maze. I watched the mate as he wove the ship back and forth between channels. After a manoeuvre that reminded me of a racing car going through a chicane, I threw him a compliment. 'Nifty,' I said.

He gave a rare smile. 'I've got more moves than a fiddle player's elbow.'

Good one, I thought; but Shackleton said it better when he compared it to 'steering a bicycle through a graveyard'. This expression came to him at the end of the 1903 season when, on Scott's orders and very much against his own wishes, he was being invalided out of Antarctica on a relief ship.

As I write, the iceplanade about us is spectacular. The dominating players are the huge, skyscraping, flat-top tabular bergs with their vertiginous, chiselled sides cleaved from the shelf. These are the great 'Kinkering Congs' of the Antarctic. Many of them are so big that their keels drag through the seabed, slowing them down to such an extent that some are stationary. But I can see changes; not in the flat-tops, but in the smaller ones, the more traditionally shaped pinnacle bergs. This is because during the storm some of them lost their equilibrium and either rolled over onto their sides or went bottom up, revealing a much gentler set of curves, a more pitted surface and a range of thin, glassy hues that dazzle the eye.

NOON POSITION: 69° 16.833' S, 012° 02.980' W

20 FEBRUARY 2019

Journey's end

Lots of sky, ice forever, very big weather and very, very cold – that's Antarctica for you. And for the last hour I have been savouring it all.

It's early in the morning and I am sitting alone in the 'monkey island', the observation room above the bridge at the very top of the ship. My bags are packed and I am ready to go.

Overnight we reached Penguin Bukta. Not that there is anything here. No buildings, no flags flapping, not even a sign that says 'Welcome to Penguin Bukta'. It's just a sort of notional spot on the ice with great views. The ship is hove to beside the low-lying fast ice at the base of a dizzyingly high cliff of frozen fresh water, hundreds of years old, that stretches off to the horizon and beyond. Everywhere we look there are massive flat-top bergs that have calved from its wall. Where there is open water it is densely splodged with so-called pancake ice, floating on the surface like water lilies. Normally these are associated with unsettled water, so they must be a legacy of the storm.

And, of course, as the *nom de lieu* implies, there are, everywhere, lots of flappy little penguins. It is searingly beautiful. There are places in the world that have a duty to remain hale, whole and eternal, and this is one of them.

And so it is all over. Feelings swell and heave about within me like the contents of our moon pool. This has been a truly incredible project but, in the end, our quarry eluded us. In the mate's brutal reckoning, it has been 'mission unaccomplished'. We have certainly taken a pummelling, but on the other hand I am much excited by the prospect of my next wreck hunt.

From Antarctica I fly to Cape Town and then back to England, where I have been given just enough time to change my socks before heading to Montevideo, Uruguay. From there I will join a survey ship that will take me back to the high latitudes of the South Atlantic, where, just a few hundred miles north of the Weddell Sea, we will resume the search for Spee's flagship, the *Scharnhorst* – a wreck for which I, together with the Falkland Islands Maritime Heritage Trust, have spent years searching.

The *Endurance* is the Ultima Thule of shipwrecks. Our chances of finding her were never very good. There was no manual for what we did. Nobody has ever attempted anything like it before. Never in maritime archaeology has there been such a daunting challenge. As Chad Bonin observed on the very first day, when we were ploughing our way out of Penguin Bukta, there was a huge Hail Mary component to this expedition.

The hardest part was always going to be getting there. But we chewed and clobbered our way through the pack until we reached those famous coordinates. And not only did we get there, we also got *down* there and actually covered the area of seabed where Worsley said she sank. This isn't the end of the story. This is really just the beginning, for other suitors will follow and one of them will succeed. This ship has a date with destiny.

The *Endurance* will haunt my thoughts for the remainder of my days and, because we did not find her, there will, I suppose, always be a small, insistent sadness within me; the baleful presence upon my mantelpiece, the bullet in my gut. But nobody can take away what we achieved. We kicked open the doors, we paved the way, set the table and even plumped the cushions for whoever comes next. And whoever it is, as I said earlier, I wish them well. I only ask that they find the wreck within my lifetime, because I am consumed by curiosity about what that ship looks like.

The question that really wriggles and claws about within me is this: did we find the *Endurance* without knowing it? The story of what happened to AUV 7, and the answer to my question, is all there within the little vehicle's databank. One day, I know AUV 7 will be found, and when that happens the information should still be intact, so hopefully they will be able to extract it. The whole situation reminds me a bit of Jules Verne's Captain Nemo telling Professor Aronnax that their story will be placed in a water-impermeable zinc container and sealed within an unsinkable casket, which will be thrown overboard by the last survivor of the *Nautilus* so that it might one day wash up upon some distant beach and the truth of what happened be known.

The mate came down from the bridge. Throughout my youth – by which I mean the 1950s and early '60s – *Reader's Digest* ran a series of biographical articles entitled 'The Most Unforgettable Character I Have Ever Met'. I hadn't thought about them in decades but, as I watched the mate approach from within the hangar, the memory suddenly reappeared in my head.

We shook hands. 'Take care,' he said.

'Don't go bumping into any icebergs,' I responded.

I climbed up beside the pilot. He handed me the headphones. I turned my head towards the side window and, as the helicopter rose, looked out over the glacier to the snow-clad, unnuanced uplands of the continent behind. In the foreground, a little behind the edge of the ice cliff, I could see the old Second World War twin-prop Dakota that would fly us inland to Wolf's Fang. Below, with the nose of her red-painted hull buried in the fast ice, was the diminishing form of the ship that had been my home for the last 50 incredible days. On the ice flats beside her were the little black dots of penguins.

Out to the west, from where we had come, was a vast landscape of great bergs with a network of deep indigo channels between.

Oh, Antarctica, I thought. You are infinitely mysterious and wonderful to behold, but I do not love you.

I took one last look at the ship, now no more than a red smudge within a white-pebbled sea.

NOON POSITION: 70° 10.302' S, 002° 07.911' W

21 FEBRUARY 2019

Cape Town

I am sitting at the bar of a hotel in Cape Town, scratching my head and wondering if the last seven weeks really did happen. I've trod a lot of snow in my time, but never anything like that. Behind the bar, a raised TV is streaming a news channel.

Chad saunters in and sits on the stool beside me. 'That sure was some rodeo,' he says, in his distinctive Louisiana drawl. For a while we just sit there in silence. Everything feels a bit aslant.

'You were lucky,' he says.

'Really?' I murmur, in a tone intended to convey irony.

'Yeah, if we had found the *Endurance* we were going to strap you down and tattoo Worsley's coordinates across your forehead.'

We are joined by marine biologist Lucy Woodall, who also looks a touch adrift. 'I have been on lots of expeditions,' she says with a shrug, 'but nothing like that.' She plonks herself down beside us. There follows a long intermission during which the three of us let our thoughts wander.

'Some ride,' says Chad at last, more to himself than to us.

Yep, I think.

And that's the way it was. Fifty-one days. A team of champions. A ton of money. The best technology. We gave it our all. We didn't quite get there. But it sure was one hell of a ride.

NOON POSITION: 33° 54.371' S, 018° 25.020' E

PART TWO

The Endurance22 Expedition 2022

31 JANUARY 2022

Cape Town

Shackleton never gave up. Nor, it seems, do we.

And so we are back.

Almost three years have passed since my colleagues and I (but especially the latter) slunk back to the UK, tails between our legs. Beaten by the ice.

I am standing on the quayside of the East Dock in Cape Town, peering up at one of the finest ice ships in the world. She's the cherry-red S.A. *Agulhas II*, the same 12,900-ton, 440-foot-long vessel we used in 2019. We might have failed in our mission but she gamely got us where we needed to be, and she's going to be our home again for the next five to six weeks as we head into the frozen cauldron of the Weddell Sea. Except we cannot go on board yet, because the fuel is being loaded for our two helicopters.

I'm joined by Chad Bonin, one of the small group who will be remote-piloting our new submersibles in the search for Sir Ernest Shackleton's *Endurance*. He was, of course, also there for our first attempt three years ago. Like me, he is delighted to be back. 'A twice in a lifetime opportunity,' he drawls.

We are the first wave to embark. There are about 10 of us. Tomorrow, the work of mobilization begins. Over the next four days, more of the team will be arriving until we number 65. If all goes well we should be leaving sometime late on 5 February. Once out of Table Bay we will shape a course for the Weddell Sea, 3,000 nautical miles to the southwest. Depending on speed and weather conditions, it should take us about 10 days to get there.

And then we will again have to shove, bludgeon and break our way through to the heart of the pack, where we will plunge

3,000 metres into the abyss in search of the most famous lost ship in the world.

I cannot promise, any more than last time, that we will receive history's tap on the shoulder; but, succeed or fail, once again I know it will be an extraordinary adventure.

NOON POSITION: 33° 54.371' S, 018° 25.020' E

1 FEBRUARY 2022

First full day back on the Agulhas II

John Shears (who is again our expedition leader) and I have the same cabins we had back in 2019. I say cabins, but they are actually well-appointed suites with sofas, tables, swivel chairs and long desks that support televisions, music players, computers and printers. Above all, they have two large square ports overlooking the front and side of the ship. From these not only can I look down on the stevedores as they load our cargo, I can also observe the crew as they stow it within the holds beneath the foredeck.

After breakfast, John and I go up on the bridge for our first meeting with Captains Knowledge Bengu and Freddie Ligthelm, our old friends. On and off, they have been working with each other for years; Knowledge tells me he has just come back from the Antarctic on the *Agulhas*, while Freddie has recently completed a stint as master of the cruise ship *Silver Cloud* out of Puerto Williams, Chile – which, curiously enough, was in the Falklands several weeks ago while I was there. 'Good to be back?' I ask.

'Definitely,' says Freddie.

'We have unfinished business,' adds Knowledge.

Freddie and I look at the latest satellite reconnaissance of the Weddell Sea, something we used to do every couple of days during the last expedition. The pack is not the solid fist of ice that most people think it to be. It is constantly on the move; leads are always opening and closing and the floes are forever mutating. Although ice coverage is densest at the centre, where the *Endurance* is located, we are hopeful that we might find ways through the outskirts that will bring us within, say, 50 to 80 nautical miles of the search area. At present there are a couple of approaches, one from the west and one from the east. As Shackleton used to say, there is no such thing as a straight line when you are navigating ice.

Freddie is particularly interested in a giant tabular iceberg that calved from the Larsen D ice shelf last June. Designated A-69, it is 19 nautical miles long with a maximum width of 10 nautical miles. Freddie is a little concerned because, like me, he remembers its predecessor, A-68 (100 nautical miles long, 25 nautical miles wide), which compromised our science programme three years ago. A-69 is currently about 200 nautical miles south-southwest of our search box and may well be grounded (i.e., have its keel in the mud), so I do not really see it as a threat, but because it might be heading in a northerly direction we will nonetheless keep a weather eye on it. Tomorrow we will be joined from Germany by chief scientist Lasse Rabenstein. He is a sea ice specialist who has been monitoring the situation in the Weddell Sea for us since November. During the search to come, his daily satellite updates and analysis will be crucial to where and how we proceed once we are in the pack.

I leave Freddie and go for a walk around the ship. Nothing much has changed except for a new lick of paint. The big difference, of course, is that Covid now stalks the corridors like some black-mantled, befanged old vampire in search of blood. The whole vessel is treated as a bubble: nobody comes up the gangway unless they have been swabbed in a small unit on dockside. Every morning we all test ourselves and then record the results on a tick-box sheet on the wall as we head in to breakfast, where we sit apart. We try to stay in our cabins as much as possible and when we go out we are fully masked. I find it absolutely stifling.

I asked our doctor, Lucy Coulter, how long she expects us to wear the masks once we put to sea. 'Oh,' she said rather breezily, 'about two weeks.' I did some rapid-fire calculations and worked out that by then, we will be on site. Jolly.

Highlight of the day? Undoubtedly it was the arrival of the first of our two helicopters, a Bell 412. At about 1530 hours, with Table Mountain as its backdrop, we all watched as it swooped in towards us. As it levelled off to land it paused above us as if for effect –

whop-whop-whop – and then landed with a kiss upon our helideck. Perfect.

We have a media team of seven on board. One of their object-ives is to make what director Natalie Hewit calls an observational-style documentary – by which I think she means quietly watching people at work while talking to them about what they are doing and what's happening within their fields of speciality.

Today, for instance, they filmed me unpacking. In particular they were interested in the charts I had brought with me in a tube. Of course, as before, one of these was the famous Admiralty Chart 4024 covering the Falklands, South Georgia and the Weddell Sea. I pointed out to Natalie that what I like about this chart is that for the area within the armpit of the Antarctic Peninsula, where we are heading, there is nothing. All the seabed contour lines suddenly cease – it is completely blank. The whole area has not been surveyed; from a hydrographic point of view it is *terra incognita*. It is where the map ends. It is where the *Endurance* sank.

Put another way, when we enter the pack, we are going off the map.

I find it really remarkable that humans have recorded every feature of the moon in tiny detail; we have a lander on Mars probing the planet's subsurface; and, as for *Voyager 1*, it is now out there in interstellar space transmitting data from beyond the Kuiper Belt. We know more about the rings of Saturn than we know about our own Southern Ocean. To me it is painfully paradoxical that although we can peer 32 billion lightyears across the observable universe, we cannot see to the bottom of the Weddell Sea.

NOON POSITION: 33° 54.084' S, 018° 25.556' E

2 FEBRUARY 2022

Mobilizing ship

The ship is a bustling hive of activity. Crew, team and subcontractors, all in orange coveralls, white hard hats and black metal-capped boots, are everywhere. Workstation containers are being winched onto the afterdeck. Tons of polar diesel and Jet A1 helicopter fuel are going into Cargo Hold 3. Above the bridge, Satcom antennae are being installed which, we are promised, will give us 'good on-the-move connectivity'. The noise is intense; the constant hammering, the crackle of handheld radios, running motors of all kinds and the weird hissing that welders make when they are burning metal. Punctuating it all is the ship's intercom, which periodically pings to life with instructions from the bridge: *'Now hear this, now hear this . . . we are pumping fuel. All smoking and hot-work to cease.'* And so on. In three days everything has to be loaded, lashed and stowed for sea. It is all about good organization and forgetting nothing; if we are going to succeed, not a single trouser button can be missing.

Earlier I passed Chad on the back deck. He was fuming; somebody had just winched one of the boxes he needed into the foredeck cargo hold. We surveyed the scene about us. I asked where our Sabertooth search vehicles were and he pointed to one of the large containers. It could be said that everything on this project comes down to three things: is the *Endurance* in our search area? Will the ice let us through? And will the Sabertooths perform to expectation?

The Sabertooths are very different machines from the HUGIN AUVs (Autonomous Underwater Vehicles) we used in 2019, which were torpedo-like in shape. When on mission we would rendezvous with the AUV periodically for a systems check but, other than that, it was fundamentally an autonomous submersible that followed a programmed set of instructions. For much of a dive, it was largely out of our control. As for the data it accrued, that was banked within the vehicle for downloading and analysis once back on deck. In other words, if there was a point of interest (POI), we would not know about it until much later. You will remember that in 2019 everything was going well until AUV 7 failed to turn up at its designated meeting point – and owing to the conditions, we eventually had to abandon the search for it as well as our expedition.

From what we learned in 2019 it was obvious that in order to defeat the pack, we needed a different set of tools. The solution subsea project manager Nico Vincent came up with was the Sabertooth, a new type of submersible recently developed by Saab in Sweden (see Figure 1 overleaf). Compared to the HUGIN it was slower and had less battery life, and was therefore limited in the area it could cover; but it had some key advantages that made it better suited to the challenges ahead. Its main asset was that it was a tethered instrument with a Kevlar-sheathed fibre-optic cable of 3.5 millimetre diameter connecting it to the ship, so that if, for instance, it suffered a major technological collapse, we would know exactly where it was and could send in the second Sabertooth to execute a recovery. The tether also gave us the benefit of being able to work in real time with the vehicle and see the data as it was coming in on a cascading sonar screen. Should a POI appear, the vehicle could immediately fly in like a drone to carry out a visual evaluation and then, if the target proved to be of significance, its side-scan could be switched to high frequency, enabling it to conduct a high-resolution sonar survey.

Almost all the team are now on board. One of the new members to arrive came by my cabin. His name is Tim Jacob and he is an educator with an organization called Reach the World, based in New York. His remit is to take us via video link into classrooms around the globe so that I and the other specialists on board can talk directly to the students, answer their questions and, in our own bumbling way, try to inspire them to go out and attempt big things of their own.

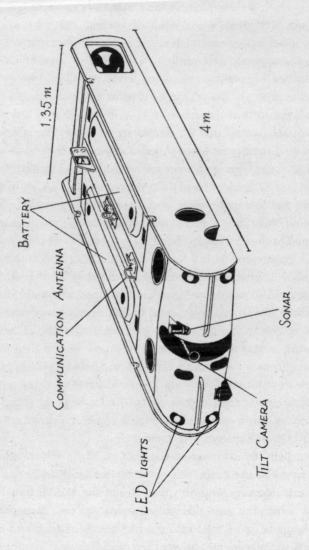

Figure 1. Sabertooth submersible used for the 2022 expedition

1.35 m

4 m

BATTERY

COMMUNICATION ANTENNA

SONAR

LED LIGHTS

TILT CAMERA

I have worked with Tim before; he is a natural teacher and kids love him. He is also a Shackleton enthusiast of the evangelistic, card-carrying, copper-bottomed kind who, before he had even accepted my offer of a cup of tea, had launched into a homily on William 'Bakie' Bakewell, one of the lesser-known figures (and the only American) on the *Endurance*. Forty minutes later, as he stood up to leave, he asserted that we were 'writing the final chapter of the Shackleton story'. Good one, I thought, and scribbled it down so I wouldn't forget.

One of the joys of this whole endeavour has been getting to know some of the descendants of the extended Shackleton family. Last August I met Pippa and Roderick Wordie over the internet, and some time afterwards Jo and I met them for lunch at a restaurant in the Cotswolds. Pippa and Roderick are grandchildren of James Wordie, the geologist on the *Endurance*. In terms of who he was and what he achieved, he was clearly one of the most admirable people on the expedition. In later life he became head of a Cambridge college and president of the Royal Geographical Society, as well as helping to establish the Scott Polar Research Institute. Meticulous, enquiring and disciplined, he was a scientist to his Scottish fingertips, so it is not entirely surprising that he was the one who made the famous drift map of the *Endurance*.

In fact, it is much more than a map; it tells the story of a broken ship upon a sea of daggers. It starts on 18 January 1915, when the *Endurance* became icebound. You can then follow her progress down to her furthest south of 76° 58', and then her zig-zag route as it is carried north on the gyre to where it sank. Thereafter one can trace the drift of the men on the floes until 9 April 1916, when they took to their boats and, six days later, made Elephant Island. The story of this document is ultimately one of hope, determination and survival, but it is also a cautionary tale: a warning that seems to say, *If you sail my icy main and presume upon my frozen core, I may crush you like a bug.*

Occasionally during our 2019 trip, some of the more nervous-natured crew members would ask me if we were safe on this ship.

I always responded in the affirmative. But when I look at the satellite data and see giant floes that weigh millions of tons, I cannot
help but wonder what would happen if we got caught in a force
10 and were nipped between two of them.

The day before I left to join the ship for this second expedition,
a package turned up on my doorstep from Pippa Wordie. Inside
was a picture of her grandfather, a photo of the *Endurance* slumped
over in the ice with a worried-looking Shackleton gazing out over
the pack, and, between two sheets of cellophane, a copy of Wordie's
drift chart. I have put them all up on my wall alongside Admiralty
Chart 4024.

NOON POSITION: 33° 54.084' S, 018° 25.556' E

3 FEBRUARY 2022

Helicopters and ice camps

Earlier there was a knock at my door and before I had a chance to open it, my visitor did it for himself. He made no attempt to enter, just hovered in the doorway in a surgeon's gown and face mask. Young, slim, tallish, dark complexion, and from behind the edges of his mask I could see what might be a thin beard.

'Hello,' he said. 'I am Shawaal. I am your steward.' I stepped forward and held out my hand. He froze. 'I must keep two metres' distance. Regulations.' There was an awkward pause. 'I have been told I must take special care of you. Is there anything you need that I can get for you?'

I thanked him, saying I was fine. Ten minutes later, he was back. This time he didn't knock, just turned the handle and came in backwards with a huge box in his arms, which he placed in the middle of my floor before smartly stepping back. 'It's for you,' he said, in a voice that suggested I might be too old and befuddled to recognize my own name emblazoned in large letters across the top.

I opened it. Inside was a whole collection of polar clothing, from socks to beanies and everything in between: coats and under-coats, salopettes, thermal underwear, snow goggles. There was even a note from the suppliers wishing me good hunting.

As I was trying on my new clothing, my handheld radio crackled to life with news from the back deck: the K-Max was coming in to land. This was our second helicopter. It is a new type of American helicopter, or synchropter, with a strange array of intermeshing rotor blades. I rushed to the hangar to watch it land. Apart from the unusual flurry of its double-rotor system, it has a narrow, wedge-shaped front to its fuselage with bulging side windows so that the pilot can look down on his load from either side. It has a maximum lifting capability of 2,700 kilograms – more than its own weight. We really need a heavy-lift helicopter and this one has been brought in at huge cost as air freight from the States.

Throughout all of our planning, the big unknown has been the ice. Will it let us into the search area and, if it does, will it give us the open water we need to deploy and recover a Sabertooth? Or will the pack be so thick and tight that we have to cut portals through it to access the sea below?

Ideally, Nico and the subsea team would like to be able to work off the back deck, using the ship's A-frame to lower the Sabertooth into the water and afterwards to raise it back on board. But – and this is Plan B – if the ice is too dense for the vessel to manoeuvre then we will have to disembark and move all operations onto the floes, where drilling teams can cut holes through which we can launch the submersibles. These sites may end up being many miles from the ship, meaning we will also need ice camps to provide sleeping and eating facilities. Above all, there will need to be a hut or shelter within which the submersible pilots, surveyors and data analysts can work. So we have brought a shelter along, made in Provence, France, mainly of metal – and although it is small, it's heavy. As indeed are the Sabertooths, their launch and recovery systems and, of course, the ice augers that will do the drilling. That is why we need the K-Max. The Bell helicopter, by contrast, will mainly be used to ferry people between the camps and the ship.

But there's a little bit more to it than that. Because we are operating a tethered vehicle with limited reach, everything will come down to drift. The pack is always on the move, but its speed and direction are forever changing. We have with us a German company called Drift & Noise, under the direction of Lasse Rabenstein, whose role will be to predict where we should locate each ice camp so that by the time it's up and running, it will be over the position that next needs to be surveyed. Because the life of each camp will be short, we will always have to have the next camp up and ready, bearing down upon where it needs to be – and sometimes, at the same time, we'll even have a third camp in preparation. There's also another company with us, called White Desert, which will erect the camps and feed the teams. Both helicopters will be running 24 hours a day, which means a doubling of pilots and mechanical support staff for each. As far as we are aware, nothing like this has ever been attempted before.

NOON POSITION: 33° 54.084' S, 018° 25.556' E

4 FEBRUARY 2022

The Falklands Maritime Heritage Trust

The last three years have been busy ones. No sooner had I returned to the UK following our 2019 search for the *Endurance* than I was back in the South Atlantic hunting for the *Scharnhorst*, the flagship of Admiral Graf von Spee's lost squadron that went down in battle off the Falkland Islands on 8 December 1914. I had made a previous attempt at finding her in 2014–15 using an old fishing trawler (that was the expedition on which we encountered the enormous wave), but after five months of being whacked about by the most consistently savage seas on earth, we had to give up because of equipment failure. This time, however, using a modern deep-ocean survey vessel and a fleet of AUVs identical to the one we lost beneath the pack during our *Endurance* search, we found the *Scharnhorst* on the Falklands Shelf at a depth of 1,610 metres. She was sitting upright, guns poking in all directions, gashed wide open by the shell fire that had sunk her. The pictures went all over the world and the Smithsonian Channel aired a documentary about the discovery, made by TVT Productions. Finding the *Scharnhorst* took away some of the sting of not having found the *Endurance*.

The search for the *Scharnhorst* had been under the auspices of the Falklands Maritime Heritage Trust, whose chair, Donald Lamont, was also part of the committee that oversaw the search for the *Endurance*. Because of our success in finding the *Scharnhorst*, the management of the new search for the *Endurance* was also passed to the Trust and its four trustees, of whom I was one. In the shaping of the new project, we decided to re-engage John Shears in his old role as expedition leader. An important addition to the team, however, was Nico Vincent from Marseilles in France, one of the best underwater engineers in the business. He and I go back quite a while. Nico used to work for Comex, the pioneering French subsea company whose founder, Henri Delauze, was a close friend of mine and a sponsor of my wreck excavations back in the '80s and '90s.

Nico is now a part of DOS (Deep Ocean Search), headed by John Kingsford, who also played a vital role in the 2019 search for the *Endurance*. On our current search Nico is in charge of all subsea operations while another old friend of mine, former French navy commander Jean-Christophe Caillens – known as JC – will oversee all back-deck activity. He and I have spent a lot of time together in the Atlantic and Indian Oceans and I regard him as a kind of mentor. Like me, he has been a professional diver and has done a lot of deep diving – in fact, he still holds the French navy's record for the deepest dive. He is somebody with a lot of wisdom to impart, and he does this with patience and generosity.

The Sabertooth is so new that there is no established pool of pilots and technicians from which we can draw the skill-sets we need. In broad terms Nico's role so far has been to assemble a team, then arrange for pilot and maintenance team training at the manufacturers' in Sweden. He also had to oversee the engineers in France who designed and made the launch and recovery systems as well as the ice-drilling augers that we will need if we have to work from the floes rather than the ship. This extremely challenging remit was not made any easier by the global pandemic and associated supply shortages, but Nico managed to get everything mission-ready on time to meet the *Agulhas II* in Cape Town. In October of last year, John and I went out to France to meet the new back-deck team where they were testing the equipment at various locations around Provence.

The team was formally introduced to us at a *zone industrielle* on the outskirts of Gardanne, a small town beside Mont Sainte-Victoire, the famous hill that was so often painted by Cézanne. Nico gave us a tour. He showed us a long, red-painted trailer, old but stylish, where he slept the team in five tiny cubicles, three bunks apiece. I asked about its unusual colour. 'It used to belong to a circus,' he explained in heavily accented but perfect English. 'Keeps them all together, you don't want to spoil them, good for team bonding.' He took us into an aluminium hut to meet the team. It was dirty, smelly and utterly charmless. Everybody was seated at trellis tables on metal folding chairs. On each table there was a plate of finger-length almond biscuits. Of the 12 men and one woman, I knew four. They were all in identical orange coveralls and hard hats.

Nico took the end of one table. He rapped twice upon its surface with a single extended knuckle. Everyone came to order. It was clear who was the boss. Nico pointed to me rather theatrically with an extended right hand. 'This,' he said, 'is Mensun. Our objective is to put a smile on his face. When that happens we know we have found the *Endurance* and made history.'

'Gents,' he went on, meaning John and me, 'meet the team. In this room we have the greatest concentration of deep-water technological talent in the world.'

Then it was our turn. John, a veteran polar geographer and expedition leader with over 30 years' experience of working in Antarctica, gave a mission briefing on Endurance22 and explained the aims and objectives of the mission, setting out the huge challenges that lay ahead for the team. Then I got up. I wasn't quite sure what I was going to say. In the end it devolved into a kind of pep talk. You know, the old *Top Gun* bit: you guys are the elite, you are the best of the best – but we are going to make you better, because this job is important.

NOON POSITION: 33° 54.084' S, 018° 25.556' E

5 FEBRUARY 2022

Back on the road

During the early evening we cast off and eased out into the steam. After all the press we received yesterday, I thought there might be a few spectators there to see us off. Not a bit of it. In fact, there were more seals in the water around us than there were people on the quay. Apart from the longshoremen, the only onlooker was a lady with two children. She cut a rather forlorn figure, and the way they lingered made me realize that they must have someone special on the ship.

When Shackleton slipped his lines at the West India Docks, London, on 1 August 1914, there were quite a few people there to see him off. A lone piper was playing 'The Wearing o' the Green' – which was also the song they were playing on the gramophone when the ice delivered the coup that finished the *Endurance*.

When the *Endurance* departed on the first leg of her voyage to South America, Shackleton was not on board. He followed some days later on a steamer, joining them in the River Plate. By not mentioning anything of this voyage in *South*, he gives the impression that he was present all along. In fact, the voyage from England to South America was highly eventful: the cat, Mrs Chippy, jumped into the ocean; there was brawling in Madeira, for which some of the hands were arrested and one man flogged; and while at sea some of the men got 'mad drunk' and one tried to throw himself overboard. They even ran out of coal, so that by the time they arrived at Montevideo they were burning the ship's woodwork.

Some have written that Shackleton stayed behind to attend to business matters relating to the expedition. Maybe, but other researchers have said that he had to make his peace with his wife,

Emily, with whom he had quarrelled – and also attend to matters regarding his brother Frank, who had just been released from prison for fraud.

One difference between us and the *Endurance* expedition is that when they unfurled their canvas and sailed off for Antarctica, they did not leave a proper support organization behind them and, although they later acquired a radio receiver, they did not have a transmitter. In other words, they were unreachable. Some have speculated that this was deliberate because Shackleton knew his creditors would soon come a-knocking. We, by contrast, are in constant communication not just with our base in the UK, but with the world. We have our phones; there's the internet; and, of course, we have a website, an educational outreach programme and, finally, a sophisticated media team fronted by the television presenter Dan Snow.

Broadly speaking, the field team can be divided up into six main divisions: subsea and back deck, science, aviation, media and educational outreach, medical, and ice camps. Each division has its own support staff working out of the UK, South Africa, France, Germany, Sweden and the United States. There has never before been a maritime archaeological project of this scale and complexity.

I have been looking at our press coverage from yesterday. They are calling it 'the impossible search'. Predictably, some have wondered if it might have been better to spend our budget on Antarctic science. This is a fair question, but it does make me smile, because much the same was said when Shackleton sailed. 'The Pole has already been discovered. What is the use of another expedition?' questioned Winston Churchill. There were even some on the *Endurance* who doubted their purpose. 'I don't see the use of Polar exploration at all,' wrote one diarist on the day of their departure from England. As for ourselves, unlike the 2019 expedition, our mission is first and foremost to find the *Endurance* rather than to conduct broader scientific research.

We have 3,272 nautical miles to go before we reach the ice edge. Right now we are bowling along at 16 knots. That is fast. The idea is to get there as soon as possible so that we can maximize our search time. If we can maintain this speed, even raise it, we could do it in 10 days.

As I write, evening has descended but we can still make out behind us the darkened mantle and distinctive flat top of Table Mountain with the city lights below. The sea is calm, the air is warm and everybody is cheerful. An hour ago we counted five breaching humpback whales. They ignored us completely; too busy bubble-feeding and getting on with their lives.

We have two people isolating in their cabins with Covid. It was a difficult decision, taken by John and Captain Knowledge, to bring them along in spite of the virus, but these are people we cannot manage without once we hit the ice. Needless to say, we are taking it very seriously: keeping our masks on all the time, and keeping our distance from one another.

NOON POSITION: 33° 54.617' S, 018° 25.591' E

6 FEBRUARY 2022

Preparations for Endurance22

This is our first full day at sea. We have unblemished skies and the ship is pounding along at 18 knots – but this is not a pleasure cruise. Today, the talking stopped and everybody began to focus on what lies ahead.

The shaping of this second search, which we call Endurance22, began almost three years ago, not long after I came back from the *Scharnhorst* expedition. The ice had given us a good kicking on our earlier attempt; personally, I felt as if somebody had walked all over me with crampons. But there was no point feeling sorry for ourselves. We had to learn from our mistakes and come up with solutions, which meant new search systems and better methodologies. In all this we had one significant advantage over last time, and that was experience. We now understood the ice pack and the nature of the risks it posed.

The engineering challenges we all faced, but especially Nico, were truly daunting. Because of Covid restrictions we were often unable to travel between England and France, so once every week or two we would meet online to review progress. I lost count of the times Nico presented the trustees with a new or evolved set of engineers' drawings, and how many times they were scrapped. In the end, though, he got us to where we needed to be, despite supply chains breaking down all over the world and businesses everywhere grinding to a halt. Things that had been simple before Covid were suddenly fiendishly difficult. From beginning to end, preparing for Endurance22 was incredibly demanding.

As I look back, there is one indelible moment that occurred over a year ago and still hangs in my mind as if it were yesterday.

We had reached an impasse. We were clean out of ideas on how to proceed. It was then that Nico said something that seemed to crystallize everything – something that was blindingly obvious, but just had to be said. In his heavily accented English, he looked his laptop square in the eye and said: 'Gents, this really is a very, very complicated project.'

Today on board has been nothing but team meetings. This morning it was all about Covid protocols, muster drills and safety briefings from the ship's officers, and this afternoon there is a marathon three-and-a-half-hour presentation about how we will proceed once we are in the search area.

How do I feel now that we are, at last, on our way? On the one hand I am brimming with excitement but on the other, there is apprehension. All the preparation work that has happened during the last 1,075 days is now funnelling down at dizzying speed to a single result. Success or failure. It is completely binary; there is nothing in between. It is either going to be cheers and hallelujahs or, once again, this whole thing is going to rise up and bite me in the bum.

NOON POSITION: 35° 20.602' S, 012° 35.313' E

7 FEBRUARY 2022

Ice conditions

We are now starting to pay close attention to what is happening within the Weddell Sea pack. Today we had our first ice strategy meeting, chaired by the expedition's chief scientist, Lasse Rabenstein. These sessions will now take place every day at 0930 hours in the auditorium. Also present will be Marc De Vos, our chief meteorologist, who will report on the weather ahead as well as that over our destination.

Today they presented the latest MODIS satellite images of the ice in the search area and its surrounds, courtesy of NASA; TerraSAR-X radar satellite imagery, courtesy of the German Aerospace Centre DLR; and optical imagery from Sentinel-3, courtesy of the European Space Agency. What these show is that the ice is currently being driven south by the wind, but that will soon change and the whole gyre will resume its northerly trend. At present, ice conditions are good. From the ice edge to our search area is 125 nautical miles, of which the last 25 to 35 pass through old, gnarled multiyear stuff that will test the ship to the limit. But of course we still have over a week to go, and in that time the pack could open further – or, just as easily, it might consolidate.

The weather this morning was perfect. The sea was calm and we have been galloping along again at 17 to 18 knots under unimpeachable skies. But there is a big eastward-heading storm a little to the south of us; the isobars are crowding, the barometer is plunging and we are promised a rough night. We practised, though, what is called 'storm routing'. These days, weather forecasting is so precise that a ship rarely needs to pass through the centre of a major gale system. You see them building several days ahead, so vessels can usually avoid their most venomous parts. On the marine weather displays we can see not only the big storm that lies a little to the south of us, but another that is barrelling in behind it from the west. Between them, there is a corridor about 60 miles wide.

Our intention is to scoot in between the two systems and thus avoid the worst of each. To do this we will throttle back to 12 knots, let the big one move a little more to the east, and then shape a course to the west-southwest that will bring us around behind it. We will still have to cross the tail of the first storm, which should mean winds of up to 30 knots and wave heights of about four metres that will slam us on the beam, causing the ship to roll dramatically. The bridge has issued instructions for everyone to lash or stow anything that might go flying.

NOON POSITION: 37° 46.080' S, 004° 34.758' E

Out of the storm

It was a bumpy night, with the ship not only rolling but also rising and slamming in a way that made everything vibrate; but it soon passed. By breakfast, a few snarly whitecaps were all that remained. By lunch we were in a heavy swell and dense clouds a little to the west of the Greenwich Meridian. During the early morning we had passed latitude 40, so we are now in what they call the Roaring Forties. This means we are crossing the 'clipper way'. In the old days, before steam and the opening of the Suez and Panama Canals, the great square-riggers on their way out from Europe to Southeast Asia and Australia used to circle around by Brazil and then, about where we are now, they would hook onto the prevailing westerlies, which would generally carry them to their destination. The further south they went, the more the winds would pick up and the faster they flew, but if they went above the forties and into the Furious Fifties they could find themselves in the ice belt. So it was a balancing act between speed and safety.

I wonder when we will see our first iceberg. When I was younger I used to lecture on cruise ships to Antarctica, and we would always give a bottle of champagne to the first person who spotted one. This, however, is a dry ship again. We do have a bar which opens every evening between 8 p.m. and 10 p.m., but it serves only soft drinks, alcohol-free beers and the usual snacks. For the beer drinkers on board, and there seem to be quite a few, this is hard.

During the afternoon, Chad Bonin and I did a live broadcast to a school classroom in England. Speaking to schools across the globe is a large part of what we now do. Reaching out to the children is one of the Trust's leading objectives. We want to use this expedition to enthuse the young, because the next generation of scientists will be critical for the well-being of our planet, and where we are heading on this voyage is the front line in the struggle to combat environmental change. Whether or not we find the *Endurance*, we want this to be our legacy.

We have now completed some 1,100 of our 3,272 nautical-mile transit to the ice edge. Those crystalline white shores beckon.

NOON POSITION: 39° 43.872' S, 002° 35.270' W

9 FEBRUARY 2022

Ice science

We are still ripping along at 17 to 18 knots in westerly seas. Despite the very long duration of this trip and the fact that we will be deep within the embrace of the Weddell Sea, it occurs to me that we will not, at any point, be seeing land. This voyage is from Cape Town to Cape Town.

There is only one thing in this world worse than gardeners talking about manure, or knitters comparing knitting, or indeed archaeologists like me mulling over stratigraphy, and that is ice scientists discussing ice. So when I see them all seated together at dinner, I usually leave them to it; but today I was late and there was nowhere else to sit. And guess what? I enjoyed myself. They asked after my health, we all expressed our astonishment at the lovely weather we have been having . . . and then they got down to discussing ice. But here's the thing: it was actually really interesting and in the end I came away feeling quite repentant. So much so that I decided I would attend a presentation they were giving this afternoon in the wet lab on Level 3.

The master of ceremonies was Stefanie Arndt, a sea ice physicist from the renowned Alfred Wegener Institute for Polar Research in Germany. She gave a show-and-tell talk on the equipment they will be using. Their main instrument (at least to my mind) is one for measuring the total thickness of snow and ice, from the surface to the ice–water interface below. This will be vital information, both for the safe landing of helicopters and for building ice camps on the floes. We have to be certain that they are sufficiently thick and stable to sustain our activities. The other tools and instruments Stefanie displayed included various drills and something called an

SMP (SnowMicroPen, or high-resolution snow penetrometer), which is driven into the floe and gives you the density and other physical properties of the snow that covers the ice. Then there is the ice corer, used to extract long cylindrical samples in which the scientists can identify different types of ice as well as determine their temperature, salinity and porosity.

As for Stefanie – or Steffi, as we call her – when she is on the floes we will not see much of her because she will be head down in a freshly dug snowpit, analysing the vertical snow structure in terms of its temperature, density and stratigraphy. She will seek to identify the different layers of the snowpack, from which she will then sample single snow grains in order to record their precise shape and size.

NOON POSITION: 43° 41.639' S, 008° 10.345' W

10 FEBRUARY 2022

Life at sea

There is little to report. We are still racing south by southwest through unruffled seas at a highly respectable 16.7 knots. Temperatures are down to single digits and I notice that more and more people are wearing their expedition-issue polar coats when they are out on deck. Nobody seems to have noticed that yesterday we passed the midpoint in our transit.

Our lives have slipped into a series of routines. One meeting follows another and in between, for John and me, it is mostly admin; I have never known a project to generate as much paperwork as ours. We are up every day at 0630 hours. The first thing we do is self-test for Covid and then at 0700, coffee in hand, John, Nico and I meet with Captain Knowledge on the bridge or in his office on Level 8. This is 'the captain's meeting', during which he briefs us on what's happening on the ship and whatever we are doing is run by him for approval. At 0730 we have breakfast and then at 0800 the Expedition Management Team meets to discuss progress and any issues that have arisen. We are nine in all. Apart from John and myself there is Nico Vincent (subsea), JC Caillens (back deck), Carl Elkington (ice camps), Natalie Hewit (media and education), Lucy Coulter (medical), Lasse Rabenstein (science) and Michiel Swanepoel (aviation). It's all pretty lacklustre stuff right now, but it will start to get interesting the closer we get to the ice. Today, for instance, the only matter of contention was what to show next in the auditorium on movie night.

At 0930 we head for the ice meeting to see the latest satellite data from Lasse, find out from Marc what the weather is brewing, and hear Captain Freddie's views of how things are unfolding within the pack. This meeting can run to almost 1030, by which time we have only one hour to head back to our quarters to attend to paperwork and emails. At 1130 we all converge on Deck Level 4 for lunch, which we gobble down fast because there are two sittings, each of 30 minutes.

Our afternoons are equally full. The last meeting of the day is between John and me in my day room at about 2200. It isn't a meeting in the formal sense, just something we have fallen into; a moment to relax and put on the kettle I brought with me from England. We go over the events of the last 16 hours, have a laugh, reminisce, talk about family, etc. We tend not to discuss the next day, because it's going to be at our throats in just a few hours. It's a kind of an unwinding thing, a little like the *amen* cadence that you get at the end of a hymn.

NOON POSITION: 48° 12.259' S, 014° 47.244' W

11 FEBRUARY 2022

Yesterday's talk of meetings and how we manage ourselves got me thinking of the way it was on the *Endurance*. Shackleton kept things simple; maybe the odd ad hoc meeting with Wild and Worsley, but not much more. Basically, he kept the team in the dark as to his intentions but sometimes gave the impression that they were involved. His team was chosen carefully so as not to include any, in his words, 'hard nuts' who might stand up to him. Despite the veiled criticisms of him expressed in the diaries, the only one to actually rebel against his authority was dear old Chippy, the hard-working, crusty carpenter from Glasgow whose rebellion during the march from Ocean Camp was something that Shackleton never forgot or forgave.

On this project John urges us to be as open as we can with everybody, and although there are matters we cannot discuss, by and large we succeed. I have been directing large-team archaeo-logical expeditions for well over 30 years and during that time I often thought that I was not so much managing archaeology as managing egos. I cannot pretend that there were not strains in 2019 but, so far, on this expedition everybody seems fairly content within their various domains and areas of expertise.

At this morning's ice meeting Captain Freddie told us that, in theory, we could be at the ice edge in three days. He stressed 'in theory' because there are three weather systems ahead of us and he is not certain that we will be able to dodge them all. Once we pass the South Sandwich Islands and are into the Scotia Sea, we can expect to take a biffing. Most importantly of all, he said there is a deep corridor into the pack which, if it does not close, will cut at least a day from our passage time. As for the search area, the floes there are still reasonably loose and dynamic; but of course all that may change and the ice could yet solidify into an impenetrable shield.

In the afternoon there was a talk by the head of the White Desert team, Carl Elkington, about what to expect if we have to move operations from the ship onto the ice. He discussed ship-to-ice radio communications, helicopter management, the clothing that must be worn, what will be provided within the tents his team will be erecting and, of course, how we will be fed. Our drinking water, he told us, will be melted snow. Sunscreen and dark glasses must be worn at all times to avoid snow blindness. Everybody, of course, wanted to know about the toilet arrangements, which will be basic. There will be procedures that must be followed. Guys, for instance, 'must pee into the blue bucket'. And when it is all over, everything, but I mean *everything*, must go back onto the ship.

Later, there was a storm warning from the bridge. I happened to run into Freddie, who nonchalantly predicted: 'It's going to hit us on the beam, and she'll be rolling like a pig.' Because of the position of the South Sandwich Chain and our urgent need to get on site as soon as possible, this is one we can't dodge.

By the evening we were being sideswiped all over the place. In one particularly gut-wrenching shove the ship heeled 30° to starboard, sending everything on my desk flying. A cupboard door sprang open and everything spilled out. My room is now a shambles. There's no point in picking anything up as it will just end up back on the floor again, so the carpet is a sea of books, papers, clothing, cups, plates and food and one football. There are 40-knot winds and seven-metre waves expected tomorrow.

But there is some good news. The two members of the team who had Covid when we left Cape Town have today been released from isolation. We are a closely knit group, so it feels strange when new faces suddenly appear in our midst. I imagine the men on the *Endurance* must have felt a little like this when they discovered the stowaway Perce Blackborow, who came aboard in Buenos Aires and would later prove invaluable as a steward.

NOON POSITION: 52° 28.642' S, 021° 32.163' W

12 FEBRUARY 2022

'We fired a Hjort mark harpoon, no. 171, into a blue whale'

I'm feeling good. Last night I solved a little twister within Shackleton's book *South* that has been chafing away at me for a great many years.

I love one-liners. 'Madame, I never eat muscatel grapes' from *The Count of Monte Cristo* is a good one, but the best cut-you-dead one-liner has to be God roaring at Job from out of a whirlwind (when the latter was whingeing on about how badly he had been treated), 'Where were you when I made the earth?' Unfortunately it's hard to find a good one-liner in Shackleton's writings, which is surprising for somebody who was quite splashy with his words.

But there is one. In the middle of a paragraph in *South*, and apropos of absolutely nothing at all, he writes: 'We fired a Hjort mark harpoon, no. 171, into a blue whale on this day.' And nothing more.

What does it mean? There is not a single clue in the paragraph and it has bothered me for decades. About five years ago, I was at a meeting with the directors of the Sandefjord Whaling Museum in Norway. These were the best whaling historians in the world. I asked them what it meant and not one of them had the least notion.

Anyway, last night at 0500 we passed the live volcano of Zavodovski Island at the westernmost tip of the South Sandwich Chain, which connects underwater with South Georgia to the west. I got up early and went onto the bridge hoping for a glimpse, but it was dark and sea conditions could not have been worse; the ship was buried in mist and the island about 12 nm [nautical miles] away to the east. I could see nothing. Back in my day room, there was no point trying to sleep, so I reached for *South* and started thumbing.

Suddenly, there was that line. 'We fired a Hjort mark harpoon, no. 171, into a blue whale on this day.'

And then I had an idea. From what he said, the date was clearly 18 December 1914. Why not see what Wordie had to say on the same day? Thanks to the kindness of his descendants, I had recently acquired a full, unedited copy of his diary. So I looked up that date – and, sure enough, there was the answer! He wrote: 'A new amusement was started this evening – firing short 1 ft arrows at passing whales. The date, etc., is scratched on the arrow just before it is fired off from the gun. The objective of course is to learn something about whale migration if possible. The arrows came by the "Corunda" from Norway, sent out by Dr Hjort.'

His mention of arrows then reminded me of something Orde-Lees had described in February of the following year, by which time they had become beset in the ice pack. It described how Wild, probably out of curiosity, had used an arrow to kill a seal rather than use his rifle, as was his custom: 'Wild shot the old Weddell [seal] with a brass arrow shot out of a rifle – an outfit supplied to us by some Norwegian balaenologist for the purpose of marking whales by shooting the brass arrows, should we have an opportunity of so doing. Each arrow is numbered. The effect of shooting the arrow at the seal was that it passed clean through its skull, and was lost in the snow. It was probably a unique way of killing a seal.'

At breakfast I could not suppress my elation; I just had to share my little triumph with those at my table. Apart from one or two little coos of make-believe interest, they were sublimely incurious. Sometimes scholarship, like virtue, has to be its own reward.

The rough weather persists, with wave heights of six to seven metres, but it is not as bad as we had feared. We have altered course slightly so that we are taking it over the quarter; in other words, it's coming towards us approximately from behind. I keep thinking of the tiny

James Caird making the crossing from Elephant Island to South Georgia in these very waters.

This evening there was a showing of the documentary *Endurance: The Hunt for Shackleton's Ice Ship*, aired on the History Channel following the 2019 Weddell Sea Expedition. Afterwards there was a question and answer session chaired by Dan Snow, involving myself, John and the four other veterans of that expedition on the ship.

NOON POSITION: 57° 39.918' S, 029° 08.298' W

13 FEBRUARY 2022

Crossing the circumpolar current

We are now well within the Scotia Sea and during the last 24 hours we have navigated our way through several boundaries, both notional and real. We have crossed the 60th parallel, which means we have left the Furious Fifties and are now in the Screaming Sixties. Temperatures have dropped perceptibly and the wind is both fast and relentless.

If I headed due east from here, I could go right round the world and end up back in the same spot without once having touched terra firma. In these latitudes it is just one unending expanse of ocean, which allows the wind to gather 'fetch' without meeting any barriers. We are also facing into the circumpolar current, the dominant circulation system within the Southern Ocean that flows clockwise from west to east all round the Antarctic Continent, keeping the warm water out and the cold in. It is the current that creates the Weddell Sea gyre that causes the ice to be constantly on the move over the *Endurance*.

And finally, we have passed the Antarctic Convergence and are into a marine belt of varying latitude where the cold, northward-flowing Antarctic waters meet the warmer sub-Antarctic waters. Where they make contact there is a sharp water temperature change which, if you are travelling north–south, can see a drop from 5.5°C to less than 2°C. The upwelling and mixing where they conjoin creates a zone that is nutrient-rich in krill and other zooplankton. This naturally attracts seabirds. Since the South Sandwich Islands, Shackleton had been seeing Cape pigeons, terns, mollymauks (black-browed albatross), nellies (giant petrels), sooty (albatross), wandering albatross and whalebirds (prions). Today we have seen

more of all these birds than before, and I have also spotted royal
and light-mantled albatross.

The first encounter with the ice was off Zavodovski Island
when, during the night of the storm, I counted 15 bergs on the
radar. As morning broke, we found ourselves completely buried in
mist and surrounded by Cape Horn 'greybeards'. Occasionally we
could just make out the spectral forms of nearby flat-top and
pinnacle bergs. On the bridge all binoculars were straining for any
growlers that might have crept in under the radar.

Today we have been seeing bergs everywhere. Interestingly, it
was after passing through the South Sandwich Group in December
1914 that the *Endurance*, just a few miles from where we are now,
also had her first confrontation with the ice. Shackleton wrote:
'The presence of so many bergs was ominous, and immediately
after passing between the islands we encountered stream ice.' By
evening, 'the ice was grinding around the ship in a heavy swell'
and Shackleton was becoming concerned.

Today, nine of the back-deck team go on shift so as to be in a
12-hours-on/12-hours-off routine by the time the real work begins
and the cold descends in a couple of days. Shifts will run from
12 to 12. Everyone is now starting to gear up for operations.
Helicopter crews are conducting emergency response drills. The
scientists are busy preparing their equipment. JC conducts checks
on all Sabertooth payload systems and reports to the management
meeting that both machines are 'dive-ready'. The only thing he
cannot do is wet-test their buoyancy; for this we have to pause ship,
but it will done before we enter the pack.

All non-essential people are now prohibited from work areas.
Everybody checks their radio comms and grab-bags in case of a
rapid scramble. Mealtimes on the *Agulhas* have previously been
signalled by a xylophone tinkle over the speakers from the bridge
but this has now been stopped so as not to wake those sleeping.
The best news of the day is that the most recent radar and optical
imagery coming in from Sentinel-1, Sentinel-3 and other polar-
orbiting satellite systems confirms that the corridor through the

pack in the direction of the search area is still available and, if anything, has even opened slightly. It now has a fairly sharp bend in it, for which reason Lasse Is calling it Neptune's Finger.

Also today, I heard the first rumblings of discontent. Some of the team who work within the close confines of the tiny back-deck shelter, as well as some of the helicopter crews, say they cannot wear facemasks at work any more. The doctor, quite properly, says that we should not unmask before our fifteenth day out of Cape Town. This is difficult for her; clearly she sympathizes with those struggling to wear masks and carry out their operations, but as she points out, if we unmask now and a Covid outbreak occurs it will spread like wildfire – which could be catastrophic for the project just as we are about to start the search.

NOON POSITION: 62° 40.082' S, 031° 00.874' W

14 FEBRUARY 2022

Into the Weddell Sea

During the night we slipped into the Weddell Sea. There was no fanfare, nobody was told, and when I mention it to people at breakfast they are surprised.

When the *Endurance* entered, Shackleton saw it as 'a gigantic and interminable jigsaw-puzzle' of ice. For us there are odd bergs and a few lumps and small floes but the sea is still relatively open so, while there is still good visibility, we can pound along at close to full speed. We could be at the ice edge by midnight. Everybody is busy doing fire and safety drills, risk assessments, toolbox meetings, group briefings. Within the team there is a sense of what parachutists call 'ground rush', when everything seems to be coming at you with breakneck speed.

I used to think that the entire 20th century could be found within the clarinet glissando that marks the beginning of Gershwin's *Rhapsody in Blue*. Everything is there: all the pain and all the glory; you feel the good but along with it is all the stupid stuff, the shaming, the benighted, the crooked and vile; but also there is everything that is weird, wacky and wonderful. All in a single clarinet call. But last night something happened to change my mind.

The thought of being back in the Weddell Sea kept me awake, so I went up on the bridge. There were only two officers and a couple of cadets. Except for those in the engine room, the rest of the ship was asleep. At night ships always turn off their lights on the bridge. This means there is no reflection from the windows, which of course helps you see out into the blackness. You sit there, alone in the dark with your thoughts, looking out over the bow into a vast ocean you can barely see. Your feelings swell and heave with the vessel. To me it is almost spiritual and, if ever I doff my agnosticism, I know that it will be on the bridge of a ship at night that I find my path to God.

So, there I was in the dark and all was quiet. But then from somewhere at the back of the bridge rose a song, I think from a phone within somebody's pocket. It was barely audible and it caught me by surprise and went straight through me. It was the rasping harmonica wail that prefixes 'Love Me Do' by the Beatles and somehow this has now become my defining song of the 20th century.

It got me to thinking about the importance of music on the *Endurance*, and the role it played in keeping the men alive on the ice. The power of music to lift and sustain was something that Shackleton understood and when he was selecting his team he sought, where possible, to recruit men who could carry a tune. Recalling his interview, the physicist Reginald 'Jimmy' James later expressed his surprise at not having been questioned about science. 'All that I can clearly remember,' he later wrote, 'is that I was asked if I had good teeth, if I suffered from varicose veins, and if I could sing.'

So important was music to Shackleton that, on the day when they abandoned the *Endurance* and everybody was discarding their possessions, he would not let Leonard Hussey leave his banjo. Hussey described how Shackleton came up to him on the ice and said, 'I've just been back to the ship, I was in the wardroom – it's a frightful mess, the beams are snapping like matchsticks, but in the only corner unharmed I found something of yours.' It was his banjo. Hussey doubted the wisdom of taking it but Shackleton insisted, saying it was 'vital mental medicine'.

Again and again in the diaries there are references to Hussey leading their frequent singalongs on the ship or, when they were on the ice, going around the tents regaling them and lifting their spirits with the old favourites. He never stopped. He even played 'It's a Long Way to Tipperary' to the penguins. Orde-Lees wrote of him, 'He is a banjoist of unusual merit, and it is very pleasant to have music of any kind down here. His repertoire is sufficient to prevent his tunes becoming too monotonous.' Today Hussey's old 'jo is on display in the National Maritime Museum at Greenwich.

Music was everywhere on the *Endurance*. The bosun John Vincent sang shanties when they were hauling in time on the ropes ('Blow the Man Down' was a favourite); when Worsley and Orde-Lees came back to the *Endurance* in a boat towing two dead seals (one of which they had slaughtered by cutting its throat), the men on board started singing 'See the Conquering Hero Comes'. On another occasion, while out on the ice, Shackleton and Worsley danced a one-step while somebody sang and whistled 'The Policeman's Holiday'.

And then there was a lovely account given by Wordie of how they used to sing after lights out: 'Saturday's concert has not been without its effect, and almost every night there is rowdiness of some sort after "lights out". On Monday it chanced to be Empire Day, and we accordingly sang mockingly all the songs once popular during the Boer War. But tonight we went still farther, and went on singing in the dark till after 11 p.m. The Boss came down, and before we ended there was a general "tuck-in".'

Wild appears to have been the best singer, with what one diarist called 'a fine bass voice'. Marston was another with an esteemed set of pipes and Third Mate Alf Cheetham also came in for praise. Tom Crean, by contrast, who much enjoyed singing, received mixed reviews. Shackleton wrote of their boat journey to South Georgia in the *James Caird*: 'One of the memories that comes to me from those days is of Crean singing at the tiller. He always sang while he was steering, and nobody ever discovered what the song was. It was devoid of tune and as monotonous as the chanting of a Buddhist monk at his prayers; yet somehow it was cheerful. In moments of inspiration Crean would attempt "The Wearing of the Green".'

We even know which songs the best singers favoured. Wild, for instance, loved 'Forty Years On' and 'Ford of Kabul River'. Cheetham sang 'Teddy O'Neil' and 'False Flora'. Frank Hurley couldn't stop singing 'Waltzing Matilda', and as for Marston, he would take on anything. Even Chippy got in on things with 'Robbie Burns' and 'March of the Cameron Men'.

As for musical instruments, there was a mandolin in the fo'c'sle, at the front of the ship, which appears never to have left the crew's quarters, and chief engineer Lewis Rickinson had a fiddle on which he occasionally accompanied Hussey. Orde-Lees also brought a banjo with him from England, but was so outplayed by Hussey that he never dared produce it.

And then, of course, there were the gramophone concerts they held every Sunday evening. If ever we find the *Endurance*, I can't help wondering if we will see the gramophone? It is there somewhere. Not to mention all the records.

NOON POSITION: 66° 10.333' S, 043° 23.511' W

15 FEBRUARY 2022

Arrival at the ice edge

'I have been thinking much of our prospects,' wrote Shackleton. And so too have I. Shackleton's words allude to his feelings as he was just about to leave the ice; I am thinking them just as we're about to enter.

We have arrived at the ice edge. Do we, this time around, have an appointment with history? I have no idea, but one way or another, once we enter that pack, it will – for me – be defining. Whatever unfolds in the coming few days, when we leave, my life will either be made or broken.

It didn't matter how I spun it at the time: 2019 was disappointing for all, and there was no hiding from it. It was all over the media. When I left the pack last time I had been thoroughly whupped by the ice and, in our retreat, we had been forced to abandon our multimillion-dollar autonomous submersible. I dreaded returning to London and facing those who had placed so much confidence in me. I had a vision of being handed a pistol with a single bullet in the spout and then being conducted to the library to do the honourable thing. But when I got back, the senior figure behind the project said simply, 'Don't worry, Mensun, we are not giving up.'

And so I have been given a second chance, which I did not expect. But now, as I look out over the pack, I have little control over what happens next. As before, it will all largely depend upon the performance of our technology, the caprice of the ice and, of course, whether I am right in where I believe the wreck to be situated. And once again there are only two outcomes: success or failure.

It feels to me at this moment as if the remainder of my life depends on the events of the coming days. The others on the team are all young or middle-aged; whatever now occurs, their lives and careers will go on. At 69, I am too old to rebuild. If what happened last time happens again, there will be no third chances. Looking at the pack through binoculars from my forrard window, I know I am staring at my fate.

Yesterday, shortly before mid-evening when we reached the ice edge, the skies were overcast. There was no wind, nor any swell, and the sea was a dark indigo. Everybody was happy and I suppose I was too, but I couldn't help feeling a touch apprehensive. I know I have said this before, but this really is the kind of place where anything can happen – and probably will (and usually does).

The ice edge is sometimes a clearly defined rim, but not always and not last night. For quite a few hours we had been swatting away the tattered ends of once-mighty floes, but throughout the early evening the mix had been thickening into great wads and broad chunks hosting the odd penguin that watched us with surprise and, I imagine, not a little suspicion. Eventually we reached a point when, without saying anything, we just kind of knew we were there.

We then found a patch of open water in which to pause ship while JC conducted buoyancy tests on the Sabertooths. This was essential to ensure optimum performance under water. If buoyancy and trim are not well adjusted then the thrusters will have to work overtime to maintain pitch, roll and depth, which will be an additional drain on the batteries, and that, of course, means reduced bottom time. The Sabertooths should be about seven to 10 kilograms positively buoyant so that if, for instance, there is a power failure when in autonomous mode, they will float to the surface. From my point of view this is important because if we find the *Endurance*, we will have to go in very low both to record and to conduct a visual inspection – and if there is positive buoyancy, then the thrusters will have to be pressing down on the vehicle to hold its altitude, which means that the prop wash will be directed away from the wreck.

We have two Sabertooths. One is called *Doris* and the other, which was manufactured especially for this project, is called *Ellie*. As soon as we are in an open pool, *Ellie* is lowered into the water over the stern using the vessel's A-frame. Trim and buoyancy testing is not highly technical; it is largely a monitoring exercise in which the vehicle descends to two or three metres and is then switched to deck mode to see if it is positively buoyant by slowly rising to the surface. After that, it resubmerges and its behaviour is observed to see if the thrusters are struggling to maintain its depth and horizontal attitude.

At 2130, while the buoyancy tests were being conducted, we had our first potential disaster. Nobody saw it coming. For almost every untoward eventuality the subsea team had devised a counter-strategy, but not for this one. While the first Sabertooth was being tested, the ship's crew opened the moon pool to lower the USBL (ultra-short baseline) pole, which holds the transceiver that allows us to track the vehicle when it is on task, and also to communicate with it should its tether rupture. The moon pool has two doors that open to the sea below. To the bosun's horror, the portside leaf would not budge. With one door open, the water came gushing up into the ship as it was supposed to, but to deploy the pole, both leaves had to be raised. I talked to Captain Knowledge on the bridge. He had just been down to inspect the situation and was deeply concerned. He informed me that he was waiting for the engineer's report, but if they could not do anything the only solution was dry-docking, which would, of course, mean the end of the search before it had even begun. 'A show-stopper,' he called it.

I knew that the nearest dry dock was in Punta Arenas, Chile, five days away – and I also knew that it was booked up many months ahead. Ironically, permission had been sought by Nico to test the moon pool doors in Cape Town before leaving, but this had been refused by the ship's engineers because they did not want to let any polluted harbour water into the ship. There was a little bit of recrimination until somebody pointed out that if the problem had been identified in Cape Town, rather than take risks it would just have meant dry-docking immediately, the consequence of which would have been to roll the whole project over into next year.

The bosun and ship's engineers held a hurried meeting with the captain during which it was decided that they would close the working door and then pump the pool dry, so that the engineers could climb down into the well to get to grips with the situation. Because of the threat of rapid flooding, this would be dangerous work. There were three volunteers: the ship's senior electrical engineer, Orlando February, and two engine room engineers, Mark O'Reilly and Kobla Dlamini.

Later, after it was all sorted, I talked to Orlando, whom I knew well from our previous voyage. He said that the source of the problem had turned out to be a hydraulic locking pin that refused to budge and had to be knocked out. It was not a sophisticated repair job, but one requiring successively larger hammers. They started off with a carpenter's hammer and when that wasn't big enough they sent for the blacksmith's block hammer, and when *that* didn't work they lowered down a sledgehammer. That worked.

By 0215 the engineers had the moon pool doors working as normal. The USBL pole was lowered, testing was completed and at 0224, to the cry of 'God for Harry, England and Saint George', we stormed into the pack. By 0315 we were slashing our way through our first ice field. Captain Knowledge thinks that with any luck, we could be in the search area by late tomorrow. The last bit, he says, depends on the thickness of the ice, and could be a slog.

It wasn't a very reassuring start. I ended my report to the trustees with the words, 'Basically crisis averted and all good.' To my wife I wrote, 'Everything feels like it is converging at dizzying speed.'

Before heading off to bed I checked my messages. There was one there from Stephen Scott-Fawcett, head of the Sir Ernest H. Shackleton Appreciation Society:

Happy Shackleton's birthday – just think, Old Shacks 148 today.

I had completely forgotten.

NOON POSITION: 67° 25.962' S, 047° 49.057' W

16 FEBRUARY 2022

First dive

At the 0700 captain's meeting this morning, Knowledge informed John, Nico and I that we had made better than expected progress overnight and that we were only 50 miles from the search area. The speed with which we've arrived here from Cape Town has been remarkable. We have been flat out the whole time and have won two or three days, and those extra days could make all the difference in the quest ahead.

We are now well within the inhospitable embrace of the pack. Three years ago we had to kick and cudgel and 'blow the bloody doors off', but this time we are following a corridor that could not be better situated. It is not so much a corridor as a wedge-shaped cleft with a bend in it – Chad likens it to a giant funnel channelling us towards the *Endurance*. To me it feels as if we are being drawn in by forces beyond our control, but he insists it's a good omen and that the wreck fairy is smiling upon us. We'll see.

'Does that mean we'll find it this time?' I asked.

'Hell, yeah,' he replied. Famous first words.

None of this is to say that the corridor is open water. It is white-licked with light floes and small ice fields, all of first-year vintage, that we just plough through at seven to nine knots without deviation. It seems as anarchic and random as ever, but something is different; it is not as muscular as last time and nothing like as threatening. In fact, it's beautiful.

And we are back in the land of the penguins. Many of the floes are platforms for goggle-eyed little Adélies; they either stand there transfixed or, when one of them becomes spooked, they all scamper off as fast as their little legs will carry them.

As for our search area, that reflects the design from last time, although we have moved it slightly. My 2019 search area was determined on the basis that we had suffered a number of time-consuming setbacks so by the time we reached the site, we had limited hours of search time available. The 2019 search box was a trapezoidal area of 108 square nautical miles. The new search area is 13.9 by eight nautical miles, which we've adjusted to 382 square kilometres (148 square miles).

All along I've been confident of our two lines of latitude demarcating the north and south sides of the AUV search field. From my study of the problem I can't see how the *Endurance* could be outside those lines. But I was, and still am, less sure about longitude, which, as I've previously mentioned, is fundamentally a measure of time between Greenwich Mean Time and local mean time, and thus depends on accurate chronometers. Shackleton's chronometers had last been rated in Buenos Aires and were now no longer reliable. Which was why Worsley and James had been trying to determine the degree of chronometer error by lunar occultation (i.e. the eclipse of stars by the moon) using positions that were tabulated in their nautical almanac.

Ideally our current search will be centred on Worsley's coordinates and then, staying within our lines of latitude, we will work east and west, and then the same again, until we find her or time runs out. But it will not be as simple as that. Because of drift and ice conditions we will be unable to work in a neat, methodical manner. Instead we will follow a pattern determined by the vagaries of the ice.

The Bell 412 helicopter has a range of 60 miles, so this morning, being well within that limit, we dispatched it on a reconnaissance mission over the search area so that John, Carl, Lasse and the other scientists could examine the floes and determine which might be suitable for ice camps. Also, they planned to leave an ice buoy, which every few hours would transmit environmental data as well as its GPS position, allowing us to track the speed and direction of the ice in that area.

Things have become serious again and there was none of our usual banter. In the quarters next to mine is Charlie Tait, one of the chief Bell 412 helicopter pilots. This morning when I ran into him in the passageway, he looked troubled and said he was worried about the weather. Apparently there was 98 per cent cloud coverage, and overcast conditions like that are not ideal for helicopters on ice because there are no shadows on the ground, which means they can't easily identify ice ridges and hummocks.

The Bell 412 took off at 1130. By 1600 it was safely back on deck. The ice buoy had been successfully deployed and John and Carl reported that there were numerous floes within the search area and its southern approaches which were suitable for ice camps.

In the meantime, the ship was doing better than expected. Certainly the ice had become denser and there were places that required ramming and some tight zig-zagging to pick up leads, but it was still mostly thinnish first-year ice that we sliced through with ease. We had to wallop our way through the last few miles but we reached the search area just before 1400. There was no jubilation, just a sense of relief that we had got here and a recognition that what lay ahead would not be easy. As the ice pilot, it was Freddie's call as to where we made the first dive, and the nearest patch of open water was towards the southeast corner of the search area. Conditions were good; drift speed was 0.3 knots. The vessel made the final transit to the open patch and took up in drift mode, engines running.

In broad terms, the plan was that once the bridge had given the green light, JC's team on the back deck would launch over the stern and *Ellie* would begin its descent. Checks would be run at 1,000 metres on the vehicle, and again at 2,000 metres. Over the seabed it would go into hover mode and then, following further checks, the search would begin. On this particular mission it would pursue two lines 1,500 metres apart on a heading of 010 degrees north, following the drift of the ship. Its bottom time would be about six hours, or nine hours deck to deck. If all went well it would then take three hours to change batteries and go through

pre-dive checks before commencing the next mission. In the meantime, the ship would relocate to a position up-current of the next
dive position and then drift down onto it. At least, that was how
we hoped things would go. It was, of course, a shakedown dive
for the vehicle, but it was also a learning exercise for the team as
well as training for the bridge, who by trial and error had to develop
a method of working with the drift.

Sure enough, dive one did not go to plan. The launch at 1726
was flawless and the descent was good, but then, at 2,900 metres,
there were pressure-related problems with one of the main thrusters
and the dive had to be aborted. At 2200, John and I were on deck
to watch the recovery. It did not go well: the tether became badly
entangled around some nearby ice. For half an hour they tried to
disentangle it, but in the end about 250 metres had to be cut and
hauled back on board.

We returned to my room for a cup of tea. Any way one looked
at it, dive one had been a disaster. John was slightly more upbeat:
we had, after all, learned what not to do. And we had set the record
for the deepest Sabertooth dive there had ever been. But I felt this
was of little consolation.

NOON POSITION: 68° 27.348' S, 052° 24.337' W

17 FEBRUARY 2022

First successful dive

Overnight, the subsea team replaced the thruster and painstakingly reconstructed the cut fibre-optic cable. The vehicle was relaunched in the early morning, but there were problems with the winch and then, at 1,300 metres, they lost communications. The descent continued for another 200 metres until, at the halfway point to the seabed, the vehicle went into hover state, which is one of its emergency response modes. Once more the dive was aborted. By now, Nico and JC had been up for more than 24 hours and were exhausted. The mood at breakfast was subdued and one or two at my table were just a little stroppy. One of the subsea team dryly observed that if any of this had happened on ice, we would be in trouble. This was true.

At 1342 the vehicle was back in the water and dive three was on its way. At 1420 it was at 2,000 metres and everything was registering normal. At 70 metres above the seabed it levelled off and finally, after three years of planning, the vehicle was switched to survey mode and the search was on.

I talked with Chad, who was at the sticks. Although delighted with the way the dive was going, he was concerned about the tension problems we were having with the winch. When not piloting, he is one of several people responsible for its care and the smooth spooling of the cable on its drum.

What was most pleasing about this dive was the quality of the imagery cascading down the sonar screen. The bottom-land beneath us could not have been more perfect for scanning. It was flat, featureless and even. Basically, we are carpet-bombing the seabed with sound, and it is the reflection of that sound, or what we call 'backscatter', that allows us to determine the nature of the territory below. Different seabeds respond in different ways, depending on the hardness and texture of the bottom. Rock, for instance, returns a hard signal; silt returns a weaker, softer signal. Very often we get mixed returns, which can be difficult to interpret – especially when you are searching for a shipwreck that may be broken up and eroded, within a mixed physical environment. One thing that's been worrying Chad and me was that there might be rocky outcrops and a complexity of other natural features; in particular, we've been worried about drop-stones from icebergs. We have previously collaborated on deep-ocean projects off South Georgia where, in places, the seabed has been difficult to read because of the ambiguities caused by the random nature of their fall. We've been concerned that it might be the same here, but judging by this dive, it was virtually free of irregularities.

'Not exactly what I would call a target-rich environment,' Chad said with his usual slightly lopsided grin. The subtext here was that it didn't look as if we were going to end up wasting valuable search time evaluating bogeys. If we passed over the *Endurance*, she would stand out like the proverbial sore thumb.

The only down side to the day was that the loose mosaic of ice over the search area had tightened up somewhat and our current neighbourhood had become a solid *mer de glace*; but it was all relatively thin and without pressure. My fear was that if conditions changed we might find ourselves within the ice, like last time, in which case I wasn't sure how we would respond. If we'd learned one thing during the past two days, it was that if we were going to work from the ship then we had to be able to manoeuvre with some freedom.

Another delight of this day has been the wildlife. We have seen many small groups of moulting Adélies and this afternoon three emperor penguins walked right up to the ship for a closer look. And, joy of joys, a minke whale popped up within a stone's throw of our back deck. It doesn't get any better.

NOON POSITION: 68° 39.442' S, 052° 17.258' W

18 FEBRUARY 2022

Ship handling

Over the past two days, concerns have been growing about the state of our new winch. The problems are mainly to do with tension. The spooling of the tether on and off the drum has to be perfect and tension must always be even so that none of the lines in the row above can pull down into the row below. If that occurs, the fibre-optic core within the tether's Kevlar sleeve can rupture – and indeed this has been happening.

At the captain's meeting this morning, Nico announced that to address this issue they are planning to bring a back-up winch from the foredeck hold around to the rear deck as a replacement. This might take six or more hours and subsea ops will therefore have to be suspended while the substitution is taking place. Then he said something more worrying, which was that the replacement winch is identical to the first and, like the first, is brand new – therefore he could not guarantee that it would behave any better.

We were at the time lodged in a large floe, so all the scientists as well as the helicopter crews and off-duty ship's crew were allowed onto the ice for science and recreation. There were light SSE winds, the temperature was about –5°C and there were intermittent snow flurries. John and I got out for a while and walked a short distance to look at some scruffy Adélies moulting in the lee of some ice hummocks. I had my football with me and we decided to organize a game. As on the *Endurance*, it was mainly crew versus scientists and, following Shackleton's example, I was in goal.

So far I have focused on the subsea side of operations, but the bridge has also been on a steep learning curve. They've had to find a way to exploit the drift and position the ship in such a way as to create a pool at the stern through which we can safely dive the submersibles while protecting the tether from rogue lumps of ice. It is an art form which, in recent days, they have mastered.

The problem is that if they position the ship in open water they have no control over any vagrant ice that, sooner or later, will collide with the float that shields the tether where it enters the water. This float has been designed to roll with the punches, but if a really large chunk of ice crosses the stern it could still rupture the tether. The solution is to drive the ship deep into a floe, creating a patch of open water at the stern that is protected from rogue ice by the lane just cut. They need to find a floe that is not too thick, but can withstand a clobbering from the ship without cracking in two.

Even then it is not so straightforward. Once Freddie determines where he wants the ship to be, he has to work out the best way to get it there. As both he and Frank Worsley used to say, there's no such thing as a straight line when you are navigating the pack. To help him, Freddie has two sources of intelligence. The first is the ice projections provided by Lasse and his associates. There is always an element of uncertainty with these, but as the days have gone by and they've become better acquainted with behavioural patterns within the pack, they've got better and better at their craft. It reminds me a bit of Worsley's sextant navigation, which was part science and part art.

Freddie's second source of intelligence is, of course, the satellite imagery. With this data he is never working in real time, but as long as the lag between the receipt of images is not too long and the drift speed does not exceed, say, 0.2 knots, he can navigate the leads and pools around the floes with considerable confidence.

Ideally, he wants the ship about three cables up-current of the designated launch site so that they can drift down upon it. Once a suitable floe has been selected the *Agulhas* will drive into it, about a ship and a half to two ships' length. They will then let the vessel fall back about 30 or 40 metres, so as to ensure that we don't become wedged in if there's a sudden temperature drop and everything congeals around us. But also, by reversing a little, they can wash away any chunks that might have gathered under the hull and might then float up into the ship when the moon pool doors are opened, preventing the deployment of the USBL pole.

Once the vessel has fallen back into its resting position the engines will be put in sea mode (that is, on two rather than four engines) and run at about 10 to 15 per cent ahead, or with about 700 kilowatts of power on both shafts. This is just enough to keep a 50-metre pool at the stern clear of ice, but not enough to disturb subsea operations.

When I talked this over with Freddie, he confirmed that he had been learning on the job. 'There's no manual to consult for what we are doing,' he said, 'but I tell you something, we could not have done any of this if we were in ice like we experienced in 2019. Not only could we not have pushed in, but also, the pressure would have closed the pool at the stern.'

Over tea in my cabin this evening, John and I discussed how incredibly lucky we have been with the ice and speculated on whether it will continue. We are in a totally different world from the one we encountered three years ago. Lasse stopped by; he is delighted with the data he's receiving from the ice tracker buoy. It has helped him improve the accuracy of his drift predictions, upon which we depend when planning dives.

Statistically, today is when Weddell Sea ice is at its least ebb. One report states that sea ice mass for this region is at its lowest since records began. This might be good for us and our expedition but, of course, it is terrible news for the environment.

NOON POSITION: 68° 40.340' S, 052° 18.505' W

19 FEBRUARY 2022

What would Shackleton make of all this?

For the first time, everything appears to be running perfectly. Thanks to the historian and television presenter Dan Snow being on board, we now have a media following that amounts to many millions around the world. During the evening I reported to the followers of my chosen media platform that:

> We have just had the best search-and-survey day so far. We had one long dive starting at 0630 hours this morning. We have repositioned the ship and now as I write we have the Sabertooth back down at 3,000 metres and on task.

The difficulties we had on the first dive were forgotten and our concerns are now on the winch systems and tether management. With the Sabertooth running so well, optimism is booming, especially as tonight we will be passing over Worsley's famous coordinates for the sinking.

Personally, I do not believe that the wreck is slap-bang on the spot where Worsley declared it to be, but, paradoxically, I will be relieved when we have surveyed the area defined by his coordinates. Where we dive is not determined by us, but rather by the ice, and if Worsley's sinking spot crusts over and we are not able to bring it within our search, and if we do not find the wreck in the time we have left, then everybody will be pointing the finger at me and asking why I did not look where the captain of the *Endurance* said it was. I have been in such situations before and it is not nice.

Our main woe today has had nothing to do with operations but rather with our diminishing supply of fresh water. The ship needs to be steaming to make water and, because we are mostly stationary, we are replacing only 20 per cent of what we consume. We have been told we must keep showers down to five minutes or more restrictions will follow.

For the first time, I felt a rising tide of optimism within the team and indeed the whole ship. There is no doubt that we got off to a poor start with the dive problems we had, but everything at last seems to be going right. *Ellie*, the Sabertooth that is doing all the searching, is performing well; the seabed is ideal for side-scanning; the ice has been lenient, our satellite imagery is excellent, and our drift predictions are better than expected. Above all, rather than fight the ice, we have learned how to work with it. Rather than trying to impose our will on it, we are allowing it to impose its will on us.

NOON POSITION: 68° 42.140' S, 052° 26.388' W

20 FEBRUARY 2022

Endurance *found?*

A little after midnight. I had just concluded my diary with the words, 'We have now covered a bit more than 13 per cent of the search area and everything for once is going smoothly. If this lasts, there is a chance we may succeed.' I put the diary away, little knowing that in the shelter on the back deck, events had just taken a swerve. Dame Fortune (as Shackleton's men called her) had made her entrance.

At exactly eight minutes past midnight, as the Sabertooth was gliding along 70 metres above the seabed, it passed an anomaly 655 metres to starboard. The new watch had just come on, and most of the old watch were still there. It didn't have much height above the seabed but its configuration, in acoustic terms, was clearly that of a wreck. There was an immediate rush of excitement but, after only three hours of searching, the batteries were dangerously low and the dive had to be concluded quickly without further recording or investigation.

There was now a chain of command to be followed. First JC had to be notified; then, depending on his judgement, Nico would be summoned, and if he agreed with JC that there was a high probability of the target being a wreck, then John and I would be called; and if I thought we were in a position to declare it a wreck, then John, Captain Knowledge and I would start calling London.

The first thing I knew was at about 0025, when there was a rap on my door. This was too late for a caller and it couldn't be the bridge, because they would have phoned. It could only be news from the shelter, the small metal hut welded to the back deck, where the pilot, surveyor and data analysts sit during operations.

I knew a dive was in progress. For them to call at this time, it had to be either something calamitous or brilliant news that could not wait till morning. On the one hand, they might have lost the submersible. On the other . . . could this be the moment I had been dreaming of?

I opened the door and there were Nico, JC and two of the media team. Nico and JC had exaggerated grins on their faces. I knew instantly that they had found something truly exceptional. 'We've got something important to show you,' said Nico, who is always straight to the point.

The reason for my optimism was that JC is a placid and collected guy who doesn't easily register excitement, but right now he couldn't stop smiling. I could tell that he was confident they had found a wreck, and the only wreck in these parts was the *Endurance*.

With John, we all headed down to the back deck. When we got to the shelter, both shifts were waiting. Fred, Jim, Frédéric, Maeva, Kerry, Clément, Greg, Chad, Robbie and Joe. François, or 'Fanche', was also there. He was a former French navy sonar specialist who was considered one of the top four in the world. If I looked up to JC for his sonar expertise, then JC looked up to Fanche for his. Like JC, Fanche was a reserved and thoughtful person who did not easily express emotion, but from the look on his face I knew he was of similar mind to JC.

The shelter was tiny, because it had to be small enough to sling under a helicopter if we ended up working from ice camps. We all crammed in and they replayed the scan for me. Immediately, I also had no doubt that we had found a wreck.

The sonar image was incredible to see for the first time, but to an untrained eye it would have been unintelligible: just an abstract splat on the seabed, a bit like the blob of a trodden snail on the path to your front door. It was about the right size for a wreck, and it contained some longish components that did not suggest nature. Most decisively of all, it had evidently been dumped on the seabed from a height because it was surrounded by a clear, unambiguous impact crater.

There were only two possibilities. First, drop-stones from an iceberg; second, a wreck. I have found drop-stones before, and this was nothing like them; it did, however, resemble deep-water shipwrecks I have seen. There were two lobe-like fields to the deposit, recalling wooden wrecks JC and I had surveyed that had opened up on impact. I was puzzled, however, by the apparent absence of a debris field; and by its lack of height, as I had been expecting the *Endurance* to be at least in semi-intact condition. I had, however, no doubt that this was man-made, and the only man-made object of this size beneath the perennial Weddell Sea pack was the *Endurance*. We punched the air, embraced, whooped and clasped hands in congratulation.

While we were reviewing the data, word had rapidly spread throughout the ship and when we left the shelter the back deck was crowded. There was laughter, cheers and yowls of delight. Dan Snow was there, his eyes gleaming; I knew what this meant to him. He was somebody who lived for history, and now he was part of it. We shook hands and I said how glad I was that he was there to share it. But there was work to be done. We allowed ourselves a few minutes to shake hands and slap backs but then JC gathered his guys to prepare for the recovery of the submersible, which was expected on the surface at 0315.

On the bridge, Knowledge, John and I made the necessary calls to South Africa and London to say that we had found the *Endurance* and then we went into a huddle to decide what to do next. Fully half the ship was up and buzzing. We asked ourselves what those asleep would want and we all agreed that they would not wish to miss out on the excitement. The discovery was announced over the intercom and everybody was informed that there would be an all-team meeting in the auditorium in 30 minutes' time. At the meeting, which was also attended by the crew, everybody was shown the sonar scan and informed that we had made history. In addition we told the gathering that the wreck was only 460 metres to the south-southeast of Worsley's position. Everyone clapped and cheered. We also requested that nobody inform their families or friends, nor post anything on social media, until the Trust had prepared an official news release.

After that there was nothing for it but to celebrate, which some did all night. The only ones unable to participate were the subsea team, who were still on watch. Our doctor, Lucy Coulter, has also declared us 'disease-free', so no more daily testing and, above all, no more face masks.

I have had some happy moments at sea, but never before have I even come close to seeing jubilation like this. It was as if the whole ship, crew and team, had just erupted in a spontaneous collective show of unbridled delight. Apart from the joy of discovery, maybe feelings were intensified by the fact that we have all been caged up for days on a ship in what is generally regarded as the most isolated and dangerous part of the world. Maybe even Covid and the two years of misery that it has inflicted upon the world played a part. Whatever it was, it was utterly infectious and seemed to scoop everybody up in a single wave of emotion.

As the sun came up, I stood on the deck in the biting cold. It was snowing heavily. This is such a weird and secretive place. 'Antarctica, I love you,' I thought as I looked out over the ethereal white wasteland.

While the rejoicing continued in the lounges, JC, Frédéric, Fred, Joe, Maeva, Kerry and Clément worked to recover the submersible. They then changed its batteries, performed the pre-dive checks and had it back in the water by 0930 – but almost from the start, there were problems with the winch. Ninety minutes later, it was once more recumbent upon the deck.

The engineers went to work on it. Temperatures were brutal. Even fully layered and with two pairs of gloves, I could not stay outside for long. The meteorologists were warning of a further drop to −15°C and below. In these impossible conditions, the back-deck team had to unspool and re-spool several thousand metres of cable.

In case this happy day ever came, I had brought with me a rubber stamp which said in block capitals half an inch high

'ENDURANCE FOUND!!' Suddenly everybody was wanting me to stamp their books, maps and pictures of the *Endurance*. Several even wanted their foreheads stamped so that they could take selfies to send home.

The previous day, *The New York Times* had written a long piece that began with the words 'The Hunt Is On'. Somebody suggested we let them know that 'The Hunt Is Off'.

NOON POSITION: 68° 39.225' S, 052° 24.395' W

21 FEBRUARY 2022

Disappointment

Yesterday, during all the euphoria, a giant floe of five by eight nautical miles came bearing down upon us. Normally this would not have been an issue, as we would simply have shifted operations to another location within the search area where the ice was more hospitable. But now we are pegged to a point on the seabed which we believe to be the *Endurance*. We need three full dives to record the wreck and explore its surrounds; and only then can we all head back to Cape Town with feathers in our caps. In the meantime, we have to sit tight and take whatever the pack sends our way.

Over the course of the day, the giant floe moved in and smothered our target area. This meant that if the dive was to proceed, the ship would have to carve a lane into the floe in order to create a protected area off the stern in which to dive the submersibles. The bridge was not thrilled with this idea, especially as the sensor that hangs from one side of the bow to measure ice thickness was registering 2.5 metres, minimum. Freddie did not like it one bit, but there was no choice. It had to be done.

Around lunch today, when it was time to reposition the ship, it simply would not respond. We were royally stuck. Clearly the plummeting temperature had consolidated the broken ice around the hull. Chad and I ventured out onto the observation deck beside the upper lounge for a closer look at the floe that's now threatening to become our worst nightmare. It is patchy, gnarled, cragged and criss-crossed with pressure ridges. Clearly it has been around for a long time and seen a lot of violence.

'A mean-looking son of a gun,' concluded Chad, and then he added more thoughtfully, 'We might have a problem here.'

'Nothing a flame-thrower can't fix,' I replied, remembering something the old mate said to me three years ago when we were in a similar situation. But truth be told, I shared his concern.

Soon Captain Knowledge was swinging the 16-ton ISO tank of Jet A1 helicopter fuel from one side of the ship to the other in an attempt to rupture the fetters that held us. The situation was potentially serious; we were stuck in a multiyear floe, unable to move forward or sideways, or to retire. If the lane we had cut to get in were to cement behind us, we could become icebound. 'What if this doesn't work?' said Chad, echoing my thoughts.

But then, to everybody's relief, it did. Sometime after 1300 the vessel began grinding her way astern with Freddie conning the ship by radio from the back decks, something I had never seen him do before. We reversed out of the floe and into a lead, where we came about, and then we were on our way to where we needed to be for the next dive.

There was not far to go and by 1420 we had the Sabertooth back in the water. Communications with the USBL pole were established, and at 1433 the all-clear was given to begin the dive that promised to be the most exciting of my life.

The Sabertooth left the surface horizontally but within a few metres had assumed a vertical attitude and was heading nose first for the seabed, where it arrived just under an hour later at 1531 with its battery at 85 per cent. Low-altitude, high-resolution side-scan and multibeam surveys were commenced. The first objective is always to secure the acoustic data, but at about 1915 there were a series of data recording failures, so at 2006 we transited to the east end of the site to begin the visual survey.

By this point, the subsea team had become a bit puzzled by some of the high-frequency scans they were seeing. These were not what they had been expecting at all. As they moved in towards the site they were all on tenterhooks . . . and then, over the next few minutes, everything came crashing down. It was one of those roll-your-eyes-in-disbelief, bury-your-face-in-your-hands moments.

I arrived at the shelter and reviewed the video with JC. What we were looking at was not the hull of a ship, but rather disjointed elements of it. It was indeed the *Endurance*, but it was far from what we were hoping. There were long shapes in the mud, one of which appeared to be rounded, suggesting part of a mast, a yard or a boom. The other components were also long and may have been from the vessel's longitudinal structure. Everything was blanketed in a fairly thick covering of silt. In one place we could see wood showing through, and there were several stones that had cleaved in a way suggestive of coal. I caught JC's attention. 'Debris,' he said in a low, flat voice. This was extremely frustrating; yes, it was the *Endurance*, but it was only a minor satellite deposit. Hopefully the main body of the wreck was not too far away; but how far away, and in what direction?

One by one, everybody left the shelter until it was just myself and Pierre, one of the nicest people I have ever worked with. He looked sad. He blinked at me through his round-rimmed glasses. 'Sorry, Mensun,' he said.

For a while I just stood there, trying to think of what to say to the documentary crew who I knew were waiting outside the door. It reminded me of how I used to feel at childhood birthday parties when the music stopped and I was without a chair.

Later on, John and I talked. Neither of us could conceal our disappointment. We discussed having an all-team meeting but there was not much stomach for it, so we decided for the moment to disseminate the news by word of mouth, at least until we had some processed imagery to show. I think all of us at the top of the team felt a bit foolish. I know I did. You could almost hear Shackleton laughing.

I ran into Chad in the locker room and showed him a photo of the site that I had taken on my phone. He looked at it with a doubtful expression. 'I hope the Trust is not planning to release a splodge like that to the public,' he said slowly. 'People expect to see a proper wreck. That looks more like something a camel might leave behind.'

I headed for my office to write an email to Donald Lamont,

chair of the trustees, who was about as excited as we had been and had been busy preparing official notifications and press statements. On the way I passed through the lounge, where some of the helicopter crews and a few of the scientists were trying to relax. There was an atmosphere of high anticipation. They knew I had been in the shelter inspecting the site. I walked quickly and tried not to catch anybody's eye. I didn't want to talk about it, but Charlie Tait, my friend from the next cabin and the sharpest wit on the ship, did. He put up his hand and waved it about, like a kid in class desperate to go to the little boys' room. Nobody can ignore Charlie. 'Well?' he demanded.

I thought about it for a moment. 'Gentlemen,' I said, 'the search is back on.'

There was an awkward pause while I tried to think up something positive to add.

'Boy,' said Charlie, 'that sure killed the mood a bit.'

Back in my cabin, Shawaal bounced in. Since that first awkward encounter we've become much more at ease with one another. 'Hey there, man, find any gold?'

I replied that it wasn't really that kind of a wreck. 'But you would tell me if you did find some, right?' he asked. I assured him I would. 'So what did you see?'

I explained what we had seen, and that it wasn't what we were expecting, and that everybody was disappointed.

He nodded. 'But it was still the *Endurance*, right?'

I told him to hold onto that thought.

NOON POSITION: 68° 43.839' S, 052° 22.665' W

22 FEBRUARY 2022

Could the wreck be under the mud?

Nothing surpasses the moment of great discovery in archaeology; conversely, nothing hurts quite as much as when you have that moment ripped from you. Magicians tell you that all magic is about misdirection and deception, and so it sometimes feels with wreck-hunting. The evidence had played us false, but never had I been trounced as badly as this. As for the mood on board, it could not be lower. 'With jubilation like that, there's gotta be a lead balloon,' said Chad.

An all-team gathering in the auditorium was scheduled for 1100 today. I didn't want to talk to anybody beforehand, so for most of the morning I just sat in my cabin. John began the meeting by appealing to people not to post anything on social media about what had happened, and then reminded everybody of the fresh water situation on board and finally that we were in temperatures of −11°C and that the ship's meteorologists were predicting another severe drop. I then spoke about the wreck-that-wasn't. I tried to appear composed, but inside I was roiling. How could I ever have let this happen? I felt as if an implied *'you idiot'* hung in the air.

In the discussion that followed, somebody suggested that maybe the wreck was there but had sunk down in the mud. I explained that this was extremely unlikely as the surface pattern was too flat for that. But also it was the wrong type of seabed. When the *Endurance* sank, there was a tangle of broken masts and spars over it and beside it. I assumed that what we had seen were some of these elements that had fallen away as the ship was sinking. Or maybe components that had got hung up on the ice when the vessel went down and were released sometime after. And the reason why the site was so flat was that once they had hit the seabed, they collapsed horizontally like a handful of Mikado, or 'pick-up sticks'.

After the meeting I called my wife, Jo, to break the news. She and my boys had been as euphoric as anybody on the ship and had been planning a special party for when Donald officially released the news. I hated the idea of disappointing them. It reminded me of those moments when the kids were young and we had to tell them the cat had eaten the hamster.

There was a long silence while she took it all in. 'You okay?' I asked.

'Yes,' she said rather sadly. 'I'm okay. It's just that I feel . . . I don't know . . . all floppy, like one of your seals.' And that kind of said it all.

NOON POSITION: 68° 39.411' S, 052° 25.804' W

23 FEBRUARY 2022

The satellite deposit

The anticlimactic feeling persists. I am less despondent now, and more angry that I have been suckered. It has been 40 years since I directed my first wreck excavation and I have dealt with satellite deposits many times before. They are normally highly interesting and an important part of the overall story. Usually if you just follow the debris trail it leads you to the main deposit, but here there does not appear to be an obvious trail. At least I know I'm in the right vicinity, but the question on my mind is, how big is that vicinity?

Quite rightly, the Trust will not be putting out a statement to the media, but naturally questions have arisen. I've summarized my views to the trustees as follows:

Regarding the deposit. My impression is that it was a major dump, but that the area around it was remarkably clear of debris. Usually I would expect to see a kind of penumbra zone of smaller bits and pieces, especially having fallen from such a height and possibly through a bit of current. For such a big area full of major items I thought it was rather tidy about the edges.

Also it seemed rather flat. If there was a ship underneath then I would have expected some kind of mound with more irregularities and bits and pieces in the near environs. And then there is the triple steam expansion engines and the large boilers and steam piping. They alone would have constituted quite a presence.

It could be a section of hull and I thought for a moment it might be part of the fo'c'sle, but neither feels quite right.

And if the ship had badly broken up then surely there
would be more debris in the area.

I know, according to received wisdom, that she was
'crushed', but she was still the second strongest timber-built,
non-naval ship in existence and I think her massive primary
structure would likely have stopped her breaking into many
pieces. If she had ruptured I would have thought this would
have happened across the engine room/boiler room, or
perhaps longitudinally as on other deep-ocean wooden
wrecks I have seen.

I thought the three or four pieces of coal was
encouraging in that they were still three-quarters above the
surface of the seabed and had not worked their way down
into it.

But I will be seeing Pierre after dinner so I can look at
everything again at my own pace.

After dinner, Pierre and I reviewed the footage to see if I had
missed anything. The thin layer of mud, however, cloaks and
confounds everything; it is like trying to read Braille with mittens
on. I saw nothing new, and remained convinced that the ship is not
under the mud.

During the day we conducted one more recording dive on the
deposit, but the fibre-optic cable within the Kevlar sheath broke
while the vehicle was at maximum depth. Fortunately it happened
near the end of the dive, so the information loss was minimal.

NOON POSITION: 68° 39.755' S, 052° 18.685' W

24 FEBRUARY 2022

Further investigation of the deposit. The invasion of Ukraine

In this morning's report to the trustees, I wrote:

> It is now 0835 GMT and the Sabertooth from the night dive
> came back on deck 20 mins ago. Nothing was found. We
> are now under way and heading to the western side of the
> area that has not been covered. Though not far, we are in
> whiteout conditions and progress may be a bit slower than
> usual. Temperatures down to −19°C.

The whiteout continued for most of the day. The snow was
mixed with fog, which made it worse. Several times they had to
shovel snow overboard from the back deck.

I had been hoping that today would be the end of the doldrums
that I sensed was beginning to lift yesterday, but then news arrived
this morning that pitched the entire ship into depression. Being in
the Antarctic, we are to a large extent shielded from the events of
the outside world, but today we heard the news that Russia had
invaded Ukraine.

It feels almost apocalyptic. The world has just come through
two years of plague and now the horsemen have brought the waste
and desolation of war to 21st-century Europe. It has had a profound
effect on the team. When John and I met for tea, he talked about
the parallels with the First World War and Shackleton's team. He
was right; for a long time it was their only topic of conversation.
Many, like Orde-Lees, who was a captain in the Marines, felt guilty
that they had not stayed behind to fight, and in his diary he
wondered if he would be reproached when he got home. Wordie

1. Icecraft. The method was for the S.A. *Agulhas II* to break its way into a floe, about two ship's lengths, to create a launch-and-recovery pool off the stern, which was kept clear of ice by prop wash.

2. Around the world people were able to follow the progress of Endurance22 through the daily broadcasts of the historian and television presenter Dan Snow.

3. The back-deck team prepares to launch the Sabertooth submersible using the A-frame at the stern of the S.A. *Agulhas II*.

4. The Sabertooth, seen here in the pool off the stern, can be programmed to perform autonomous operations or can be piloted manually from a control console on the deck. A Kevlar-sheathed, fibre-optic cable between the vehicle and the surface allows sonar specialists to analyse the acoustic data in real time as it is acquired.

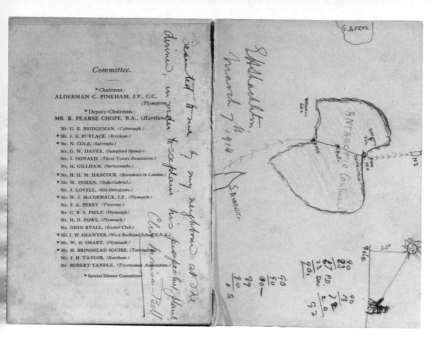

5. A sketch of Antarctica by Shackleton on the menu card of the fourth annual dinner of the London Devonian Association, held in London on 7 March 1914.

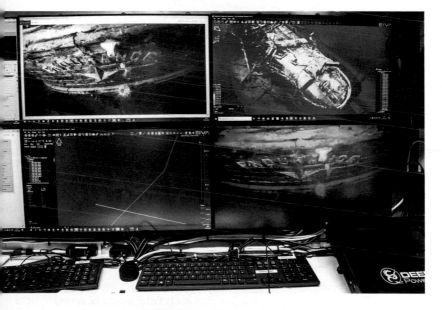

6. Screens inside the back-deck shelter showing the *Endurance* on the seabed.

7. Patience Camp. Shackleton (*right*) and expedition photographer Frank Hurley beside a stove with skis and a pole tent behind.

8. The stowaway, Perce Blackborow, hid away in a locker until the *Endurance* had left the River Plate and was on its way to South Georgia. At 18, he was the youngest on the ship. Shackleton made him their steward after promising that he would be the first to be eaten should they run out of food. On his shoulder sits Mrs Chippy, who had to be shot soon after they abandoned the *Endurance* and took to the ice.

9. *Left to right:* John Shears, Mensun Bound, Nico Vincent and JC Caillens after finding the *Endurance*. They hold a book of Frank Hurley's photographs and a laptop image of the *Endurance*.

10. A celebratory photo of the subsea team as they gather around the main Sabertooth.

11. July 1914. The *Endurance* in dry dock at Millwall, London, as she prepares for her journey to the Weddell Sea.

OPPOSITE PAGE
12. (*top*) The stern of the *Endurance* on the seabed, with the ship's name curved over the Polaris star after which she had originally been named.

13. (*middle*) The bow of the *Endurance*. The bowsprit was broken off by the ice at the point where it entered the ship. The chain from the starboard anchor (which is still in place on the fo'c'sle deck above) can be seen disappearing into the hawse hole.

14. (*bottom*) The hand rail around the well deck at the stern of the *Endurance*. Behind can be seen the intact helm, or ship's wheel, with the companionway to the cabin deck in the background.

15. An oil painting of the *Endurance* leaving South Georgia on 5 December 1914, with Grytviken whaling station in the background. Painted by George C. Cummings who was a whaler at South Georgia in the 1960s; later he became a marine artist whose works now hang in whaling museums around the world.

16. Shackleton's grave at South Georgia. The headstone was unveiled by the governor of the Falkland Islands on 26 February 1928. Rough-hewn from granite, it bears the nine-point star that Shackleton had adopted as his symbol.

was also anguished by their absence from the conflict. In his diary while they were icebound, he mentions his heightened sense of shame when he read in an old newspaper from 21 September 1914, that one of his best friends from Cambridge, R. C. F. Powell, had been killed in action. 'He was an only son,' he ended.

There is a growing school of thought on the ship that the wreck is in the mud beneath the satellite deposit. I have had deputations wishing to discuss the matter. Yesterday evening the guys in the shelter found a couple of small POIs, points of interest, which I was hoping might support my argument that the main wreck deposit is still out there waiting to be found. We took a look – and once more were disappointed. In my report back to London I wrote:

> I was called to look at a couple of small anomalies at about 10 p.m. One was a largish drop-stone. The other looked like a piece of coal. The positive I took from this was that the drop-stone, which must have been about 40–50 cm in length, did not appear to be very deep within the mud at all. You are probably aware that there has been some thought (not from me) that the wreck has sunk down into the mud; this stone would appear to argue against that hypothesis.

To me, the beauty of wrecks is that they represent a fixed moment in time. Sites on land, by contrast, often represent a great many years of occupation, and in my youth when I worked on villa sites outside Rome and elsewhere, I soon learned how difficult it was – often impossible – to assign a secure relative chronology for many of the artefacts we were finding. With a shipwreck, though, you are dealing with a single instant in which everything was pitched onto the seabed; and, because water can be a wonderful preservative, in the right circumstances the wreck and almost everything within will still be there. A wreck can be the perfect time capsule.

The *Endurance* does not appear to be a perfect time capsule. Normally the satellite deposit we found would have gone down at the same moment as the wreck, and thus the two would be close together. But if they are not near to each other, then presumably they went down in separate sinking events. So which went down first, the hull or the satellite deposit? And how much time passed between the two sinkings – and where are they, relative to one another?

From Hurley's photographs and descriptions in the diaries, it is clear that the ship was heeled slightly to starboard within her frozen cradle and that the ice had overwhelmed her from the starboard side (rather like a wave). Along her port side there was a solid ridge of ice, showing that the hull would have had to go down first, and that some of the timbers caught up in the floe went down later. In Hurley's photos and film, shot on 28 October 1915, we can see that the masts had fallen to starboard – but in photos taken on 8 November, following salvage operations from Ocean Camp, the foremast and its spars have been bundled to port and the upper main mast has been taken away. Some of these timbers were cut up for use at Ocean Camp, but what was not used would have been left lying beside the wreck. These are probably the long timbers that we found on the seabed, and the reason why they are all still together is because they were linked by their stays and other ropework. The question from the trustees and head office became: how long would they have been on the ice before they fell through? From the diaries it is clear that nobody visited the site until 24 November, three days after the sinking.

Macklin, 24 November: 'The ice [at the sinking location] had undergone considerable pressure and movement, and the pieces which had crushed her and were holding her up were separated by hundreds of yards.'

Worsley, 24 November: 'Wild, Macklin and Hurley make a trip back to the ship's last berth and find the ice that had been crushing her scattered in all directions to a distance of a hundred yards at least and nothing to indicate that a ship had been there was visible.'

Wordie, 24 November: 'Wild and a party went across to the position where the ship sank . . . Nothing whatever of the ship is left. The ice has rearranged itself considerably.'

From this it is clear that any timbers left on the ice beside the sinking hull could not have remained there for longer than three days.

Assuming that the drift was following the northerly trend, and that what we found sank a couple of days later, then the wreck must be somewhere to the south. I have checked Worsley's positions for 22 to 24 November and found that in that time, the ice drifted north from 68° 34' 30" S to 68° 28' S. If one minute of latitude is one nautical mile, then the wreck cannot be more than 9.5 nautical miles to the south.

I had a meeting with Nico before breakfast to look at Pierre's graphic for Worsley's coordinates, which I gave him last night. We also considered Worsley's observations on wind and drift. Nico and I agreed that, depending on drift and sea ice conditions, we will try to complete our coverage around where we found the deposit and then, ideally, work our way south from there when the pack allows.

NOON POSITION: 68° 42.372' S, 052° 27.028' W

Parhelia and sun pillars

We have now completed our coverage of the area around the satellite deposit. Because of adverse ice conditions we cannot move south, but the situation is favourable in the western sector, so we have repositioned the ship about 1.5 nautical miles in that direction. At 2105 hours yesterday evening, we launched the Sabertooth and managed an almost eight-hour dive in unexpectedly fierce currents of 0.6 knots.

Because of the severe low temperatures, we have been witnessing some spectacular parhelia which have had everybody out on the deck in awe. Parhelia are great haloes of light around the sun that swell into bright splurges, or so-called 'false suns', where the circle nears the horizon. They were frequently seen by Shackleton's men; on 12 February 1915, Orde-Lees wrote: 'The sun now gets very low at night . . . Two nights ago there was a parhelion, or "mock sun" at midnight. These phenomena occur when the sun is low on the horizon and the air is full of icy particles.'

Last night we saw another dazzling light display: a sun pillar, or a long, vertical beam of light that extended upwards from the sun, which was just on the horizon. As with the parhelia, this has to do with ice crystals in the air and their bending and scattering of the light. Several days ago, when the sun was higher, we saw one that also sent a beam downwards and across the ice to the ship. On 25 March 1915, Orde-Lees described the same phenomenon: 'Very bright and cold with a fine parhelion most of the day . . . about 3 p.m. there was a fine sun pillar, a shaft of bright light running down from the sun to the horizon, where it splays out into a glow of light.'

There have been vocal critics in Britain telling us that we are looking in the wrong place and that we should be searching to the east. These views have now made it into the media, who would clearly like to create a little controversy, and I have been contacted by the BBC for a response. I have made no secret of the fact that this was always going to be what is called a longitudinal search – although our lines of latitude are fixed, our longitude isn't. Ideally, ice permitting, we would begin the search at Worsley's coordinates for the sinking and then work east and west from there. In other words, we would be covering anyway the sector in question, which makes it all a bit academic. This I explained to the BBC, who thereafter dropped the matter. In diving archaeology we always work to the mantra, 'plan your dive and then dive your plan'. And that is what we are doing.

My main source of anxiety now is time. It is the first thing I think of when I wake up and the last at night. It feels like the ratcheting of a metal band about my windpipe that every day gets tighter. By the terms of our contract with the South African company that operates the *Agulhas II*, we now have only three full working days left before we must exit the pack and head back to Cape Town. There is, however, within our contract the option for a 10-day extension. John is working on this and we both feel fairly certain it will be granted. However, it isn't just the terms of the charter now bearing down upon us; it's also winter.

NOON POSITION: 68° 40.916' S, 052° 28.524' W

26 FEBRUARY 2022

'Beautifully bright but d—bly cold'
(Orde-Lees, 10 February 1915)

Recently the weather has been overcast and hostile, but today the skies suddenly opened up and everything became dazzling. The temperature rose and with it our spirits. The news from John is that the South African government has agreed to the proposed extension of our charter. I notice that somebody has built a snowman on the back deck, and today most people have been signing up for a table-tennis tournament in the helicopter hangar. Adding to the new mood is no doubt the fact that it is Saturday, which is steak night with ice cream to follow.

The scientists took advantage of the fine weather to get out on the ice and, not that they needed us, John and I went with them. A tent was set up for food and shelter. After a bit of a kick-around with my football, the two of us joined them for a meal of dehydrated food that requires only hot water from a vacuum flask to be made palatable. These were supplies that had been intended for consumption in the ice camps, but now that we have found a way to use the *Agulhas* to create a protected pool for launch and recovery operations, there is no longer any need for ice camps and so we have all this food left over.

After the meal, some of us walked to a group of moulting Adélies sheltering behind a series of large hummocks. The area around them was covered in excrement. Several scientists found the smell offensive. I told them that Shackleton also hated the smell of penguin egesta. Having spent part of my life downwind from a penguin rookery, I just love it. Why, I sometimes wonder, doesn't somebody bottle it up to make scented candles?

Back on board afterwards, I went down to the back deck and

put my head over the gunnels just as a minke whale popped up almost directly below me. His sour, salty breath caught me full in the face.

At 1530 hours Nico came to my rooms; John was already there. He told us that the analysts in the shelter had just reported two anomalies found during routine post-processing of the data. The larger is about 4.2 kilometres from the satellite deposit on a bearing of 298°, or roughly west-nor'west. The larger covers an area of about 45 x 76 metres; the other, about 270 metres away, is about 50 x 50 metres.

I went down to the operations room for a look. The anomalies do not immediately shout *wreck* but they might well be debris from the *Endurance*, which would help us establish a 'line of fall' for the sinking ship, which might lead us to the main wreck deposit. One feature was very long and narrow, the other much smaller. The long feature in particular is extremely interesting. In normal circumstance it might be a low stone ridge, but we have not seen any geology around here. Remembering what happened on 20 February, we are trying to be guarded, but I can feel latent excitement within the shelter. We will conduct an investigation during the next dive, which should be off the deck sometime during the late evening. Clearly this is going to be a long night.

There has been much discussion regarding where we should move the hunt now that we have patched in the central and western sections of the search area. We would like to cover the southern sector, but ice conditions there are still not favourable. There is, however, a large zone to the east that we have not visited and where, according to predictions, we could have two or three days of consistent south–north drift; but we will need to get across there fast, as the captain has told Nico that compact ice is moving into the area. He has also been told to expect three days of snow, which will make any ocular satellite imagery useless and safe navigation

over such a distance could become impossible. He warns that if
we do not move out of here soon, long-range relocation may
become impossible. We will therefore conduct an inspection dive
to verify the nature of the POIs and, if they are not wreckage, we
will transit east as soon as the submersible is back on deck, so as
to be within the new sector before the weather again deteriorates
and we become buried in snow.

It's now been six days since we last had an issue with the vehicle
or the winch. This is an amazing team of very talented people, but
the star in our firmament is undoubtedly Nico Vincent. I think he
might have saved the project. Back in Sweden, when he was taking
charge of the Sabertooths, he spotted an old winch in storage and
asked if he could also have that one for the project. Everybody was
of the opinion that the two new ones would be entirely reliable,
but Nico's view was that they were not tried and tested, whereas
the old one, despite its years, was. Both of the new winches have
now failed and so Nico has switched to the 25-year-old winch, and
it is working. It did not like the cold, but they constructed a tent
over it and put a heater inside, and it has performed perfectly. The
subsea team say that bringing the old winch was an act of whim,
but to me it betokens the kind of wisdom that only comes with
experience.

NOON POSITION: 68° 37.662' S, 052° 27.258' W

27 FEBRUARY 2022

The march of time

Overnight we investigated the two POIs. We carried out a high-resolution survey over the targets and then conducted visual inspections. I am afraid neither was wreckage. The one that excited us most seemed to be a long, fairly high furrow in the sediment with additional inexplicable disturbances in the immediate vicinity. The second target was also quite a distinct interference of an otherwise flat and featureless seabed, but for both targets, the conclusion was that they were natural features and nothing man-made.

Naturally we are all disappointed, but the search goes on. We are currently recovering the Sabertooth; once on board we will transit east and resume operations.

The nine-nautical-mile transit towards the opposite eastern end of the search area was challenging and required quite a bit of icebreaking. We arrived during the second half of the afternoon and by 1640 hours we had the Sabertooth down and on task. By 2240 it was back on deck; a smooth dive but it had seen nothing.

Having found the satellite deposit a little to the south of Worsley's coordinates, and given what we know from the diaries regarding the weather and their believed state of drift, I can't really see how the *Endurance* could be in this vicinity, as it would imply that she was on a strong easterly to westerly trajectory. However, conditions are not fit for the south or for the NW corner, and besides, this is wreck-hunting, a pursuit that often defies logic, so it might as well be here. Nico and I talked a bit about our current situation and how, as the search area shrinks, so do our options on where we can go next. 'It's the game,' he said with a sigh, giving that shrug of the shoulders that only the French can do.

I see myself at this moment as facing four enemies:

1. My own ignorance of where the ship lies.
2. Weaknesses within our technology.
3. The ice and the deteriorating winter conditions.
4. Time.

Regarding factor 1, the situation moves in our favour as our coverage expands. Factor 2 is also less of a concern, given both vehicle and winch are performing well. Factor 3, however, is a growing source of anxiety as each day we draw nearer to the big freeze of winter. And that leaves time, which is now my biggest worry of all.

The 10-day extension to the charter has been agreed starting on 1 March, but as Knowledge told John, and indeed as was discussed at the captain's meeting, the considered view from the bridge is that we can stay in the pack for only four or five days into the charter extension. That means we have to be heading out on 5 March.

An acquaintance in the media has contacted me to ask if the reason we have moved east is because this is where the newspapers are saying we should search. I grit my teeth and for the umpteenth time try to explain that yes, we have moved east, but it has nothing whatsoever to do with opinions that have been expressed in the papers and elsewhere. For us, everything derives from what the ice and the meteorology tell us.

Right now there is a huge multi-year floe pressing in on us from the south, and we know from satellite intelligence that we cannot fight it. Further, we must not get caught between this large floe and another big one that appears to be converging on it. We also know there are two or three days of heavy snow ahead, which will complicate ship movement as the ocular satellite data will be useless during this period. All we will have is radar and whatever cloud-penetrating satellite imagery we can download, which is rarely in real time.

The reality of this very hostile world means that we cannot go hopscotching about on the whim of armchair critics who do not understand our methodologies or the brutal realities of the conditions we face.

NOON POSITION: 68° 37.104' S, 052° 12.096' W

During the night, a twist occurred in the tether that links the Sabertooth to the ship. It didn't snap the cable, which is armour-sheathed and has a breaking strain of 400 kilos, but it did rupture the glass of its fibre-optic core. The dive was completed but data was lost and they had to perform a 'dead vehicle' recovery. Back on deck valuable time was lost trying to isolate the rent and then perform a fusion splice. All in all, it cost us several hours that right now we cannot afford to lose.

At one point Kerry Taylor, who was making the splice, looked up to me and said, 'You can watch all you like, but there's nothing that can make this process go any faster.' It wasn't until 1430 hours that we had the Sabertooth back in the water. I went down to see Chad to find out if the splice was working. He had over 4,000 metres of tether out (there is 6,800 metres on the drum) but everything was performing well. He told me the *Endurance* is so close he can smell it. I reminded him that he has been saying that since we started. 'I know,' he replied, 'but right now it is really, but I mean *really*, pungent'.

As predicted, the snow has returned. Our decks are buried and, as I write, we are in fog mixed with sleet that is so thick we can only see a few yards in any direction. Knowledge has instructed that until further notice there can be no more science on the ice. Some of the scientists think he is overreacting. Maybe, but I recall reading how members of Shackleton's team got lost out on the ice in conditions like this.

This morning Senior Meteorologist Marc De Vos sent a memo to the bridge saying that there will be a severe temperature drop

tomorrow evening. Knowledge responded: 'We will have a challenging week ahead of us, especially with the strong winds and
poor visibility.' Freddie added his concerns about wind chill factor
and what that will do to leads. As soon as he feels he can no longer
navigate safely, he 'will recommend to the Master that we call it a
day'. He also drew attention to the 'huge floe' that is encroaching
upon the search area from the SE. Freddie told me the only reason
we are here – and getting away with it – is that we are still dealing
with mainly first-year ice, and that if the ice was like it was in 2019
we would have had to leave some days ago. He thinks the reason
there's so much first-year ice is because there have been two 'good'
ice years in a row, so the backlog of multi-year ice has been
moderate. He is also concerned that with the coming deterioration
in conditions, he might not always be able to hit the launch spot
– which would mean repositioning the ship at much cost in terms
of time.

There is a sense that we are into the endgame. One way or the
other, something has to give soon.

So right now all eyes are on the giant floe. It is extremely close
and tomorrow we will have to deal with it. We won't be able to
fight it, so we are going to have to find a way of working with it.
I keep remembering how, three years ago, a 'monster' multi-year
floe pushed in from the south and ended our hopes of finding
AUV 7.

Before going to sleep, I tapped out a response to a question
from an online follower who is wondering what Shackleton would
have made of all this:

You asked what Shackleton would think if he could see
what we are doing. I believe he would share our
excitement. His endeavours were all about discovery and
adventure and, at a certain level, so too are ours. I think he
would be fascinated at the prospect of seeing his old ship
on the seabed . . . In particular he would be amazed by this
juggernaut of a vessel we have and that it could almost fit

the *Endurance* on its fore deck. And as for the technology
we have, I wonder what he would make of our
submersibles? To him they would be science fiction. He
would, I am sure, wish us well and, if we do not succeed,
he might ruefully remind us that he too did not succeed in
all the big things he attempted.

NOON POSITION: 68° 42.315' S, 052° 09.396' W

1 MARCH 2022

Within the giant floe

I am anxious about what route the giant floe will take once it is in the search area. Although the momentum of its drift is north-ward, it can change direction at any time. What will we do if it swings around, like the monster floe in 2019, and blots out the southern part of the search area where we need to be as soon as possible?

Unable to sleep, I rose before 0600 and went up on the bridge to find out what had been happening overnight. Ali, who monitors the satellite data, was there beside her computer. 'What's happening with the giant floe?' I asked.

'Oh', she said matter-of-factly, 'we are in it.'

I looked out through the wall of glass that surrounds the bridge. Sure enough, we were deep within the floe. As far as the horizon there was not a pool or lead in sight. As one of Shackleton's men said, we were 'a mere speck in a boundless frozen desert of ice'. Fortunately, it is mixed ice. In amongst the thick and gnarled old areas that we cannot penetrate, there are new fields that we can. We are also lucky that the floe is on a northward drift and although strong winds are expected soon, these will be southerlies, further propelling us in the right direction for completing our search.

The night dive, however, did not go well. An issue developed with the cable again and suddenly, without warning, the screens went black. But – and this is the truly incredible part – it happened precisely when the interrogating sonar pulses were passing over a major anomaly. The emerging feature on the side-scan cascade suddenly stopped as the data flow ceased. Within the shelter every-body's eyes were glued to the screens. Nobody could quite believe that in an eight-hour dive, this was the moment when it all went wrong. It was as if somebody had cut through the image of the anomaly with a knife; one half was clear, the other half was black.

So what can we say about the new POI? We don't know the extent of its size, but the indications are that it could be large enough to be the *Endurance*. There was, however, nothing about it that to me said 'wreck', but it was well defined, meaning it is not some illusive will-o'-the-wisp. One problem was that it did not appear to have height of the kind that might represent a ship's structure, but that could be there in another part of the feature that we did not capture. However, there appeared to be the hint of what might be an impact crater, which would confirm that it had fallen from above. Of course, it might be no more than a dead whale but, equally, it might just be our quarry.

The vehicle was back on deck by noon and immediately the subsea team set about trying to isolate the presumed rupture in the fibre optics. The plan now was to conduct a visual evaluation of the target as soon as the vehicle had been prepped and charged. However, to get back to the position over the POI within the giant floe by following a direct line through the ice will require much ramming and reversing. The ice pilot is currently thinking that it would be easier to go out and around the floe, then re-enter from the west.

The problem with the cable was more difficult to resolve than expected and the Sabertooth was not ready for launch until 1730. In the meantime, the giant floe had been moving north at a fast 0.5 knots and we had been carried with it. By the time the vehicle was ready to dive we were over the northern end of another area we need to investigate anyway, so it was thought best to take advantage of circumstances and dive there.

By mid-evening the Sabertooth was down and on task. Although the pilots were struggling with over four kilometres of tether out in unusually strong currents, they were nonetheless managing to cover terrain, albeit in a manner described to me by Chad as 'messy'.

An hour and a half later, though, everything changed. The sharp weather deterioration predicted yesterday by Marc arrived. The winds were whipping in from the south, there was driving sleet and temperatures were plummeting. Everybody was eyeing the worsening conditions with alarm, pinning our hopes on the anomaly we saw before the data flow cut out earlier.

We can't always have good days and today wasn't one of them. The best part was a moment in the afternoon when a leopard seal hauled itself out of the water at the back of the ship, the first 'lep' we have seen up close. There were emperors nearby on the ice but neither seemed to show any interest in the other. Every now and again this arch-predator would snake its way back to the ice edge, slither in, conduct a brief patrol and then re-emerge onto the ice for a snooze.

This evening I wrote to Patrick Watts, a friend in the Falklands:

Sorry I have been so quiet but the last week has been utterly relentless. I feel the pressure. As for sleep, you just sort of top it up as best you can as you go along. And then there is the cold. As I write we have −19°C with very strong winds and almost horizontal sleet. Fortunately, I am mostly inside, but for the guys on the deck it is physically sapping. We will be lucky if we can get in another five days of searching. Winter is biting and the captain is growing nervous as we are in frequent whiteouts. We will have to find leads to get out of here before the pack seals up until spring.

NOON POSITION: 68° 41.329' S, 052° 07.627' W

2 MARCH 2022

'The snow fell hissing in the brine
And the billows frothed like yeast'

—HENRY WADSWORTH LONGFELLOW
'THE WRECK OF THE HESPERUS' (FRANK WILD'S FAVOURITE POEM)

Sometimes you feel that this place wants to devour you. And so it was today.

Shortly after 0600 I drew back my curtains and found myself looking into a white maelstrom of almost-liquid sleet. It was like staring into a washing machine. I went up on the bridge to find out what had happened overnight. The giant floe is now to our north and no longer a concern but the whiteout, together with contrary currents and a gluey strudel of mushy ice, was making it very difficult for the vessel to relocate for the inspection dive on the POI of yesterday.

We knew we faced a day of savage cold and the wind chill dropped temperatures to below −30°C, the worst by far that we have experienced in our two campaigns to find the *Endurance*. Everyone on board was aware of the potential importance of this dive and, despite our expressions of caution, expectations were running high.

At precisely 0835 the vessel was over the launch site. Within five minutes of receiving the green light from the bridge, JC had the Sabertooth in the water. It was a very tough dive. Both piloting and tether management were challenging. The 3.5 millimetre-thick fibre-optic link between the vehicle and the ship bowed out in the current, puling the submersible in directions it did not want to go. At one point we had 5,200 metres of cable out in a drift of 0.7 knots and a current that wasn't much less. It wasn't until 1340 that we closed in on the POI.

Hearts sank. It was a major rock dump from an iceberg. In other words, gravel and stones that had been scraped from the ground thousands of years ago when the iceberg had been part of a glacier. Grim-faced JC gave instructions to 'end dive and return to deck'. The dismay that ran through our ship was palpable. I noticed that the sea water on the tether had frozen by the time it crossed the deck to the winch drum.

It was soon after this that one of the team came to me with a news item on his phone that – yet again – said we were looking in the wrong place and that we should be to the east; that is to say, the very area we had just completed. I pursed my lips in righteous disapproval, but frankly I no longer gave a damn. There was, however, some satisfaction in the thought that all the wiseacres in Britain who had been telling us to search that sector had been proven wrong.

Shortly after 1400 there was a meeting in my cabin to decide where to hunt next. It was a bit awkward as I had printouts from the *Endurance* diaries laid out in rows all over the carpet. There were three areas left to search but it was decided to head for the stretch across the southern side of the box. Fortunately the ice drift was perfect for a west-to-east traverse of that area.

Undoubtedly, this will be our last shot. Our situation is dire. We are in the middle of a frozen morass in raving weather. By night, temperatures tipped –40°C, which, quite literally, is sufficient to freeze mercury. Neither man nor machine can hold out for long in such conditions. But how much time do we have? Hours? One day? Two? Maybe three? And when it ends, what will we be? Victors or vanquished? Winners or losers? It is now all about holding our nerve.

At 1830 we had the Sabertooth on deck and lashed for passage. At 1850 we set off across the search area toward the southwest corner. Five hours later, a little before midnight, we were on location and ready to begin the search of the southern sector. At 0040 we commenced diving. Before falling asleep, I texted my wife: 'Let's hope the wind, current and temperature are more congenial tomorrow.'

So, what was the purpose of all those papers on my floor? Our brilliant meteorologist Marc De Vos has come up with historical weather data for the Weddell Sea that covers the period around the sinking of the *Endurance*. Working together with Lasse and some of the other scientists, he was preparing drift models for the trajectory of the ice during the critical three days leading up to, and just after, the loss of the vessel on 21 November 1915.

I confess I was sceptical when they first asked me for the weather data from Worsley and the other diarists. I had with me copies of all the diaries, which explains why I had them laid out on the floor of my cabin. I was amazed. From the blizzard of 10 November through to 22 November 1915, Marc's historical data agreed with what the diarists had recorded. There was only one day when there was not perfect congruity. I asked Lasse to write a letter to the trustees explaining the implications of their findings; I could see it was important, but it was also complex and I did not want to misrepresent their work in any way.

NOON POSITION: 68° 44.083' S, 052° 09.053' W

3 MARCH 2022

*'We seem to be drifting helplessly in a strange world
of unreality' (Shackleton, 29 March 1915)*

Elephant Island. 17 June 1916. On this day, beneath upturned boats
and by the light of a blubber lamp, all five toes on Perce Blackborow's
left foot were amputated. Frostbite, a condition that scared them
all, had turned his toes black and they were beginning to smell.

At 3 p.m. on 31 October 1915, three days after they abandoned
the *Endurance*, they shot the cat and, dragging their boats, set off
for the islands of the Antarctic Peninsula. They could each carry
1 lb of 'perquisites'. Before they commenced their trek, Dr Macklin
sewed a special pocket onto the breast of his jersey. It was closed
with four safety pins, one of which was made of gold and thus
would not rust. Into it went a few of his most essential articles.
'Two carborundum stones [for sharpening his sheath knife], a silk
handkerchief, a packet of sledging toilet paper [and] a small round
mirror.' He later wrote, 'It was good practice for men in company
to scrutinize each other's faces for frostbite' – but, because he did
not trust them always to do so, he carried the little mirror so he
could perform the task on himself.

As well as fingers and toes, their chins, ears, cheekbones and
the sides of their noses were particularly vulnerable. Although they
probably all experienced its early effects, it was a miracle that no
one else suffered like Blackborow. Lionel Greenstreet, according to
Wordie, had badly frostbitten fingertips, but recovered when the
black bits eventually fell off to reveal new but highly sensitive skin
underneath. One of the diarists wrote of how 'a finger or two got
rather badly frost-bitten, but were restored before it was too late'.

I think most of us have felt the early effects of frostbite; I have
several times. With me it is always the fingers. They burn with pain

and go white and waxy about the tips, at which point they must be quickly revitalized with warmth. The worst bit, I find, is the severe pins and needles you feel afterwards but that, I am told, is a sign that things are recovering. If you ignore the initial pain, it will go, and then you can be in trouble. Orde-Lees described the symptoms well. He had left the ship and was out on the ice when his 'toes [began to] ache with cold . . . Suddenly the aching stopped and I felt a delightfully comfortable sensation in my feet. This I knew meant that several of my toes were frost-bitten . . . I ran back to the ship.'

Yesterday winter arrived with a vengeance and overnight the wind chill again tipped −40°C. This morning we experienced the most brutal weather that we have had over both campaigns of 2019 and 2022. Conditions were extremely dangerous and everybody, but especially our doctor, Lucy Coulter, became concerned about frostbite.

For the guys on the back deck it has been truly challenging. They can only be exposed for short periods. Food, snacks and hot drinks are constantly available and the White Desert ice camp team and some of the younger helicopter crewmen have been helping to relieve them. Nonetheless, at one point the tears in one man's eyes froze, sealing shut an eyelid. I went out to watch one of the recoveries, and in that time the water within a bottle I was carrying in a bag became almost solid. It reminded me of an incident recorded by one of the diarists, when Reginald James's fountain pen froze and burst its reservoir so that, when it thawed, it created a 'grievous mess' in his pocket.

I stayed up to watch the overnight launch of the Sabertooth on its first dive in the new southern sector, and so slept a little later this morning. I was awoken when Shawaal stormed in. He never knocks.

'Hey there, old wreck dude, sleep OK?' I assured him I had. 'Bosun says we will be leaving today because of the weather.'

From my windows it looked even worse than yesterday. There

was a thin crust of ice on the glass through which I could just make out a swirling white phantasmagoria of sleet and snow. I have never witnessed such frenzied conditions as in the Weddell Sea going into winter. This place is only for the brave or the stupid, so perhaps Shawaal is correct and it is all over.

I kitted up and went down to the shelter to find out what had happened while I slept. As I shoved open the watertight door onto the back deck, there was an instant cloud of condensation as warm and frozen air met. This was something they experienced on the *Endurance*. One of the diarists wrote: 'If the door of a warm cabin is opened, the outer cold air streams in like steam, the actual fogginess being caused by the moisture in the air of the cabin condensing.'

Several members of the ship's crew were nearby. We talked a bit. I sensed it yesterday – a kind of worm of melancholy running through the ship – but today I have been feeling it even more strongly. Many people on board have had their fill of this benighted place. Just as Shawaal said, they're of the opinion that we are about to head home.

Outside, in the after-work area, the air was so thick with snow you could almost smell it. The decks were all stodgy with the stuff. The cold came at your eyes and lips like something out of a tack gun. There were only a few figures working, half covered in sleet and so bundled up you could not make out who they were. Even their eyes were goggled over for protection. Through the blizzard I could make out the oblong of the A-frame towering over them; it made me think of the gates of hell. This was real knife-between-gritted-teeth stuff. I glanced at the pool off our stern: the black water was becoming starchy with new ice. Everything felt threatening.

I tapped on the door of the shelter and went in. 'Nothing,' said one of the surveyors in response to my enquiry. They looked exhausted, which was not hard to understand. They sit here for hour after hour in this cramped, airless, windowless box, calling into the canyon for an echo that never comes.

In the constellation of beautiful minds that make up this ship, there is none finer or more faceted than that of photographer Frédéric Bassemayousse. Shackleton would have loved him; he is an eternal optimist. The underwater world is his life. Like me, he visualizes the *Endurance* as upright, heeled in the currents and breasting its way through the soft bottomlands of the Weddell Sea, destined never to make port nor ever to sight land, always the epitome of grace under sail and strength at sea. 'Don't worry, Mensun,' was all he said to me today. 'We will find her.'

I feel once again that things are slipping away from me. In the long, unscintillating history of failed archaeological endeavours there can be few, if any, to match what happened to us in 2019. And now we seem to be heading that way again. But this time it is worse. I have always said that this was an all-or-nothing project, and yet here I am in this forsaken halfway house. All we have is a few mud-covered sticks. We are the team that found *a bit* of the *Endurance*. I feel like Jacob Marley or the Ghost of Hamlet's Father, a kind of damned soul, neither in one world nor the other. Anywhere but here. For an archaeologist, that is a fate worse than death.

But we are not quite at the bitter end. I found Knowledge and Freddie up on the bridge. Normally one covers for the other, so to see them together like this was perhaps a measure of the anxiety they share. Nobody was saying much, but at least there was no talk about giving up. The Sabertooth was on deck and the ship was surrounded by ice. Knowledge was conducting ship's business on the computer. Freddie was staring out into the white oblivion, his eyes narrowed to slits like a professional darts player going for a triple twenty.

As I watched, he started nudging the ship round. He was totally focused. I was reminded of something the old mate used to say: 'Look where you are going and then go where you look.' The white grot around the hull shifted and crumpled. I could not work out what he was doing, so I asked.

'The wind just veered,' he explained, 'and I want to bring her head a few points more into it so as to give the guys on the back deck a bit more shelter.'

Chief scientist Lasse Rabenstein has written the letter I requested to send to the trustees concerning his work on hindsight drift projections for the *Endurance* during the period around the sinking. He has applied 'the same method as for the drift forecasts we use on board on a daily basis' to the 'worldwide historic weather station data' and, by aligning the calculated drift trajectories with Worsley's positions, they have been able to simulate the movement of the *Endurance* within the ice between 18 and 22 November. The results show that during the review period the *Endurance* might have drifted from the northern edge of the search box all the way to the southern edge (possibly further) and back north again. Lasse, Marc and their colleagues believe therefore that the search should be more along a north–south axis rather than an east–west one, and broadly in agreement with Worsley's positions and our discovery of the satellite deposit.

We are going to hold a meeting tomorrow morning to review their findings but, from talking it through with Lasse, I find the work compelling and all my original scepticism has expired. We are already searching the area where they would like us to be, so it will be very interesting indeed to see what, if anything, unfolds in the little time remaining.

During the afternoon there was some truly excellent news. Knowledge told me that the forecasts for tomorrow predict a period of moderate weather that will allow us to continue the search. It feels like a stay of execution. But how long will it last? Knowledge is not certain but, any way one looks at it, we have fewer days ahead than I have fingers on my right hand.

NOON POSITION: 68° 41.026' S, 052° 08.397' W

4 MARCH 2022

Sunshine and a show

I felt yesterday as if we had slipped into some kind of collective ice coma, but today the weather is sublime and that has transformed everything. Out of curiosity, I looked through the diaries to see if Shackleton's men had experienced anything as beautiful as this so late in the season.

Sure enough, on this very day in 1916, James gave a description that could have been written by any of us today: 'Have never seen the pack more beautiful. Bright blue sky and sun, but the air all a-shimmer with tiny ice-crystals falling. The snow surface seems covered with glittering sequins, and the distance is blue with a faint haze covering the bergs. Cracks and leads opening up here and there, with a faint film of ice over them, also covered with sequins and with mirror-like reflections in the still open parts. We are all agog.'

All the scientists are out on the ice beside the ship, drilling or digging ice pits. Some work alone, others in small groups. As I watched them from my cabin window earlier they reminded me of the busy little figures you get in a Pieter Brueghel the Younger painting. All so industrious, all so dedicated to what Shackleton called the creation of new bricks for the great wall of knowledge. I noticed a small group beavering away beside the starboard bow; I couldn't quite see what they were doing, so I peered closer. They were building a snowman.

We have hacked into a large five-by-five-nautical-mile floe of first-year ice that is only 1.6 metres thick. The overnight dive went well. The drift was a mere 0.1 knots and by 0820 the Sabertooth was back on deck. We have now covered over 80 per cent of the search area, including the southernmost three squares of column A. Today we are diving the southernmost ends of columns B and C, which will put us directly south of Worsley's estimated position for the sinking of the *Endurance*.

At midday I went out onto the helicopter deck. There were some light flurries, more pixie dust than snow. I looked at the screen of a drone pilot whose vehicle was so far overhead I couldn't see it. From above, we are no more than a tiny Santa-Claus-red dot on a huge white canvas. I was struck by the thought of how everything here in the Weddell Sea is either materializing or dematerializing.

Chad appeared. He had just got up to go on shift and was still a bit foggy-eyed. I chided him for sleeping away the best part of the day, something my father used to say to me on weekends.

'Maybe,' he smiled. 'But you know me, always up at the crack of noon.' For a while we just looked at the vast fugue in white about us. 'Just think what a kid with a box of crayons could do,' he mused.

We both agreed that this whole region is profoundly weird. Chad thinks it's the kind of place where spaceships come to refuel. I told him it's where snowmen come to die. In fact, the spirit of every snowman there has ever been lives on here. That's a good one, he conceded.

Inevitably, we talked about our prospects of finding the wreck in the little time we have left. Word has it that we have just two days before the weather closes; then we're out of here. Chad is confident of success; I am not. In his own quiet way, he is quite religious. He reads a chapter of the Bible every day. We have different Promised Lands, but right now I feel he is more likely to reach his than I am mine.

This afternoon, John had the excellent idea of requesting Hussey's meteorological log from the Scott Polar Research Institute at Cambridge. About an hour later, they sent through a digital copy of the original. As expected, it confirms the information in the diaries and fills in gaps. I don't think it will change the opinions expressed by Lasse in his letter of yesterday, but it will enable them to redo their calculations with a fuller, more authoritative data set and thus consolidate their findings.

In the spirit of both Scott and Shackleton, we ended the day with a concert in the auditorium organized by a young American from one of the helicopter teams, Mike Patz. He is excellent on the guitar and (like Wild and Marston) has a fine singing voice. There were some beautiful songs from both the team and ship's crew. John led with 'It's a Long Way to Tipperary' and Dan Snow gave a fabulous recitation of two of Shackleton's favourite poems, Tennyson's 'Ulysses' and Browning's 'Prospice'.

Music was very important to Shackleton's men. Every Sunday they held a gramophone concert and there were over a hundred records on board to choose from. Their favourite song was the magnificent opening aria from Rossini's *Barber of Seville*, in which Figaro storms onto the stage singing about how wonderful he is; so at the end of the evening, I gave a few words of introduction and then we played a video of it.

Apart from not finding the *Endurance*, it has been a pretty perfect day. At 2315 hours, before turning in, I went down to watch the launch of the Sabertooth. There were 10 emperor penguins porpoising about in the pool with the vehicle and half as many again on the bank beneath our starboard quarter. Enchanted kingdom above; deep secrets below. By 2330 the submersible was under and nose down, on her way. The back deck emptied. I went over to the corner of the stern for a better look at the penguins. I gazed down at them and through glassy, distant eyes they looked back up at me. Who, I wondered, is studying whom?

NOON POSITION: 68° 44.221' S, 052° 31.256' W

5 MARCH 2022

'It's the Endurance*!'*

Sometimes as life unfolds there are occasions, rare and extremely special, when everything seems to come into alignment to create a moment of absolute perfection. Spheres chime, time stands still and every strand bonds to form a single, beautifully neat cosmic bow. And it happened today. For today, you see, we found the *Endurance*.

As usual, I woke to the alarm at twenty to seven. Immediately it was upon me, the gripping sense that things were slipping away from us, and in only three days we would have to be out of this frozen barrel and on our way back to Cape Town. When you sit astride a project that costs millions, the thought of failure knots your intestines and depresses the very hell out of you.

As always, the first thing I did was go into my day room, which has views out over the bow as well as to starboard. We were still within the large five-by-five-nautical-mile floe from yesterday, but overnight we had come out of it and then hacked our way back in at a point that was well positioned for the next launch.

It was quite a heavy floe whose surface was mixed; there were a few flat paddocks of young ice, while other places were characterized by numerous long pressure ridges whose ruptured and upthrust edges had been softened by snow. Then there were areas of hummocked eruptions sticking up like broken fingers, but which also had been worn down by time and the relative warmth of the austral summer. One of the *Endurance* diarists had said that ice like this reminded him of a half-buried city.

There was, however, one standout feature to the hard-frozen acreage about us, and that was a large iceberg about a kilometre and a half to starboard. I cannot pretend that it was a particularly spectacular berg; I had seen bigger and better. Shackleton's diarists described many of the more distinctive bergs they saw and Worsley even sketched some of them. There was the 'Broken down' berg that had open archways through it, and there was the 'Steeplehat berg' that rose up in a pinnacle, and then there was 'Castle berg', which had, so to speak, high battlement towers at both ends and a crenellated curtain wall between. Our berg, by contrast, was long with raised hills near both ends and a low-slung saddle between. It reminded me of the Two Sisters that dominate the skyline to the west of Port Stanley in the Falkland Islands.

I skipped breakfast and went straight to the captain's meeting. As usual it was just John, Nico, myself and Captain Knowledge. These meetings had become a matter of routine and the only issue on the agenda today was that we were still consuming fresh water at a rate which the desalinators could not replace. Knowledge outlined the further restrictions he would have to impose. We liked none of them, but at sea the captain is God and his word reigns supreme.

After the captain's meeting there was the daily ice briefing, at which Marc gave us the latest weather forecast and Lasse talked us through the satellite imagery. There was no bad weather in prospect and the satellite intelligence told us little we did not already know. Nonetheless, it was a bit more interesting than usual, as Lasse afterwards gave a disquisition on the drift patterns for when the *Endurance* sank and presented a hindsight model on which he, Marc and another ice scientist, Christian Katlein, had been working, raising some interesting thoughts about the rough position and circumstances surrounding the ship's loss. Later in the morning we all met in the vessel's business centre to study and evaluate their findings. Their considered view is that the *Endurance* is south of Worsley's coordinates; in other words, about where we are now.

Shortly after lunch John and I were on the bridge. Both of us had been looking at the berg through binoculars. There seemed to be a small group of moulting Adélies beside it. For some time we had been talking about getting off the ship for a walk and that berg was the perfect excuse. Our charter was coming to a close and this might be my last opportunity to go out on the pack – my last chance before a humiliating return journey, with little to show for the immense expense the expedition had cost.

I am always worried about 'weather bombs', that is to say, winds and whiteouts that descend without warning. I have been caught in both, but according to the meteorologist, there was nothing ominous in the offing. John was more concerned about cracks opening up in the ice and cutting us off – which had, after all, happened to Shackleton often enough. Indeed, during their first night on the ice the men had to move camp three times because of such cracks. We agreed, however, that this was a stable-looking floe, so we decided to go for it. We figured we could be back in about an hour and a half.

Before we went off to kit up, John put out a team message:

Good afternoon All. Saturday Movie Night. By popular request of the Helicopter crews, the Saturday night movie will be the cult horror thriller *The Thing*. This film is set at the US Antarctic research station, Outpost 31, in 1982. Twelve men are working at the station gathering physical science data. It is the dead of winter. With six months of darkness ahead of them, they uncover the find of the century (and it is not the *Endurance*). If only they could put it back. Showing at 8 p.m. in the Auditorium. All Welcome.

It must have been about 1530 when we were winched over the side. We walked around to the front of the ship where Grant Brokensha, a White Desert field guide, took some photos of us beneath the bow; and then, under scowling skies, we set off for the berg. Generally the ice was firm, but in those areas where the floe was rotten its surface was like a pie crust through which, without warning, we would sink to our knees.

As we trudged along with the frozen snow scrunching beneath

our boots, we talked about how little time was left and the frustration we all felt. We had covered 80 per cent of the search area and we would not be able to cover it all in the time we had left. John is a very grounded guy, and what he said next surprised me a little. He said he could feel the presence of the *Endurance* beneath our feet. I had no doubt that somewhere in the icy-cold tenebrous canyons below lay the *Endurance*; but John said it in a way that suggested he meant *directly* beneath our feet. As we contemplated this possibility, we had no idea that events were about to take a very dramatic turn back on board the ship.

The shelter – the small, free-standing metal cabin that had been welded onto the back deck – faced the stern about eight metres away. It contained an L-shaped connecting desk that sat three people in front of three separate banks of screens. At the back of the shelter, furthest from the door, was the pilot station, from where the Sabertooth's progress was monitored and controlled. This consisted of a console, joysticks, and four large screens. The pilot at the time was Robbie McGunnigle, a big, smiling Scot from the Isle of Arran. Next to Robbie, but at right angles and facing to starboard, was the online station, where all the incoming sonar data appeared on two screens. This is where Clément Schapman, a hydrographic surveyor from France, was seated. To the right of Clément was the offline station, where the data was processed. Normally, this is where sonar analyst François Macé would have been situated, but by chance his seat was empty as he was conferring with Pierre Le Gall in the operations room. There was a third person in the shelter at that time: Lars Lundberg, an engineer from Saab, the Swedish manufacturers of the Sabertooths. On the back deck outside, Chad Bonin was monitoring the winch. He was also a Sabertooth pilot, and every now and again he would swap roles with Robbie. There were five other pilots, analysts and surveyors, but they would not be coming on shift until midnight.

From where Robbie was seated at 'the sticks' he would not normally be able to see the incoming sonar yields. At that moment, however, he was deep in conversation with Clément and facing the online screen. Because Clément's head was turned slightly, it was actually Robbie who first spotted the anomaly as it edged into view.

Everyone saw Robbie suddenly stiffen. His eyes were focused on Clément's screen, and the others followed his gaze. The vehicle was proceeding at two knots, so it was quite slow. This meant the POI was also slow to appear. All three men in the shelter watched, transfixed, as it grew in size. The first thing they saw was its outer margins, which we now know was the impact crater that the falling ship made when it struck the seabed. As they watched it take shape, it dawned on them that this was something serious. And then they began to see the acoustic shadow it was casting. That was quite big, so they knew the object had height above the seabed. This sure didn't look like drop-stones.

Robbie gave a whoop of excitement, but then professionalism kicked in and they all held their nerve and settled their attention again upon what was emerging. It dawned on them all that maybe, just maybe, this was the *Endurance*. But they kept a grip on their expectations. They all remembered how we had let our excitement get the better of us on 20 February.

In the event of any anomaly that might plausibly be a ship-wreck, there was a defined set of protocols or procedures to follow. Things had to move up the chain in a proper way, the link below deciding on whether or not to inform the link above. The first person they had to call was their supervisor, JC. Depending on what he thought, he would then call Nico, and then, depending on what *he* thought, he would call John and myself and then we would decide whether or not to call the chair of the Trust, Donald Lamont, who would make the final decision on whether this was the moment to call the authorities at the Polar Regions Department in the Foreign, Commonwealth and Development Office back in the UK.

So, following the protocols, Clément radioed JC, who was on duty. Robbie also called in Chad, who was at the winch just outside the shelter. When Chad heard the urgency in Robbie's voice, he almost immediately knew what was going on – that there was a target on the sonar scan. He opened the door and looked at the screen on the right. 'Holy cow,' he thought, and it obviously showed on his face, because when he looked up everyone else was watching him, smirking slightly. He kept his cool and told them that this was definitely very interesting, but that they needed to verify it. What he was really thinking was 'Bullseye!'

When JC received the radio call from Clément, he was in his cabin. The screen with the incoming sonar data moves with the submersible, which emits the pulse. From when a point first appears it takes about five minutes for it to disappear, and JC came into the shelter just before it had crossed the field and left the screen.

'Let's take a look,' he said as he pushed through the pack of bodies to get to the scanner. JC doesn't say a lot, but he gave them all the biggest smile ever.

He stayed maybe 30 seconds, then ordered them to stop the survey and go to starboard for a high-resolution pass on the target. He then went out and, following procedure, radioed Nico and then headed for the ops room, as the guys there would now be needed in the shelter.

Clément stopped the vehicle. It was then his job to calculate a path into the target at reduced altitude.

Nico was in his room on Level 7 when he got the call from JC requesting his presence in the shelter. There was something in JC's voice, so he grabbed his protective back-deck gear and headed straight down.

After calling Nico, JC swung the lever on the watertight door that let him leave the deck and step into the ship. Then he walked down the corridor towards the moon pool to find Pierre, François and Jim, who he knew were in the ops room a few metres away.

Pierre had just been post-processing some earlier data. A few minutes before, he had spotted an anomaly, and although it was not a major point of interest he still wanted a second opinion, so he had radioed François and asked him to join him. They had reviewed it and agreed it was not worth investigation, so together they were heading back to the shelter. As they walked along the corridor to the back deck, they passed JC coming in the opposite direction. Again, they knew from experience that JC does not give much away, but he had a broad, happy look on his face. All he said was *'Elle a pété.'* This is not polite French; it is slang and it means 'She's there. We got her.' They said 'No way' and 'Not possible'. JC just smiled and nodded in a way that told them he wasn't joking. François and Pierre looked at each other and then, together, broke into a run down the corridor towards the back deck.

In the ops room, JC found Jim and quickly told him in French, 'We have something really interesting and it's four metres above the seabed. Come as soon as you can.' And then he hurried off, returning to the back deck to wait for Nico. By then, Pierre and François were already in the shelter. When they arrived, Clément was still computing the vector we would have to follow to take us into the target, so the vehicle was still paused. François took the empty chair in front of the offline screens. At this point all eyes were on him.

Because of his years in the French navy as a mine hunter, François is the master of sonar interpretation. As far as everyone was concerned, his was the final word. He adjusted his glasses. There was complete silence. And then, without taking his eyes off the screen, he simply said, *'C'est elle.'*

Hearing that from François confirmed everything for everyone in the shelter.

The Sabertooth was still stationary, and now they had to switch it from wide-area survey into small-area target-investigation mode. Captain Freddie Ligthelm was on watch on the bridge, studying the weather data. Every now and again he would glance at the AUV screen that shows both the navigational line-plan for the mission and the position of the vehicle along it. He was a little concerned that the submersible was 1,500 to 1,600 metres from the ship, and at the current rate of drift it would not be possible for it to complete the dive. He just happened to be looking at the screen when he saw the vehicle divert from its search line, and then it went into hover mode. This by itself was not so unusual, but nonetheless Freddie kept half an eye on the screen as he returned to what he was doing. At the same time, the UHF or ultra-high-frequency radio was going off in French. It wasn't the usual tired old radio blah-blah; the tone was different. Freddie realized something was up. Somebody with a French accent was trying to radio me, out on the ice with John. Freddie turned to the chief mate and said, 'Looks like something important.'

Pierre took over from François at the offline station and began to process the SSS (side-scan sonar) file that contained the target; Clément needed this for the line-planning for the high-resolution survey. Pierre was very excited, but it was also stressful work: they had to change all the settings, and they had to do it quickly. There could be no mistakes. Everything needed to be perfect the first time because they were losing battery, there was not a lot of dive time left, and there were concerns over the drift. Besides, he knew that if he got it wrong, Nico would chew his head off.

Robbie dropped altitude from 80 to 10 metres. The sonar moved from a low frequency of 75 kilohertz to a very high frequency of 419 kilohertz and began to close on the target.

When Nico got down to the back deck, he found JC waiting for him in the freezing cold. They had a brief chat, during which JC told him they had a target, and that it should be the wreck, and they should be on it in about a minute.

The high-frequency sonar survey was already in progress when Nico entered the shelter. They had chosen a sonar range of 100 and were not flying over the target but 50 metres away from it, the target being on the starboard side of the vehicle.

At first, Nico was very calm. This time, he wanted to be 100 per cent certain.

Everyone was now crammed into the shelter like sardines in a can, pressed up tight against each other. Nobody volunteered to leave. In the entire world at that moment there was only one place to be, and they were there. As they got closer, the whole mood got more and more intense.

Robbie was at the sticks muttering 'C'mon, baby, c'mon.'

They had transited about 400 metres when they saw the unmistakeable sonar impression of the wave of mud that had been thrown up by the impact. And then the bow appeared. It was wedge-shaped and perfectly clear in outline.

Behind the bow, they could see the shape of the hull . . . and then came the masts . . . everything. What an entrance she made; it was as if she was rising up from the deep grave of history. They were awestruck.

Then somebody said, 'That's it. It's the *Endurance*!'

Those words seemed to crystallize everything. They could no longer hold back their excitement and joy. It was concrete. This time we had her. We really had her. Those who had phones whipped them out and took photos of the screen. They knew they were bending the rules, but as Chad said to me later, 'We had just found the *Endurance*; we were now the Lords of the Deep!'

Robbie went the full length of the wreck and the sonar confirmed that it was in one piece. By then they were beyond happy – they were delirious. At 1632 they completed the run. JC left the packed shelter and went back to the ops room to see Jim, while Pierre began processing the sonar images they had just captured and Clément and François built the line-plan for the next phase.

When JC got back to the ops room he quickly showed Jim the high-frequency side-scan picture they had taken. Jim was so overwhelmed that he almost felt the room spin around him. 'Sexy picture,' was all he could think of to say, and then JC headed back to the shelter. They would soon be conducting the visual verification and he didn't want to miss it.

What everyone had seen in that first pass was real enough, but it was a painting in sound, a stitched-together matrix of returning sonar pings. Now they were going to go in with both eyeballs and see it in all its physical actuality, as if they were divers. Robbie was still piloting and was adamant that the Sabertooth needed to return to the surface because of low battery. But we needed the footage and Nico insisted the dive continued. They went in at just two metres of altitude.

When they arrived at midships on the port side, they turned so that the Sabertooth was facing the ship at a distance of 50 metres. This was a real-time visual inspection: they were about to become the first humans to set eyes on the *Endurance* in a hundred years. As Chad put it later, it felt as if they were travelling through another dimension, a dimension of sight, sound and mind.

For the first 40 seconds they saw nothing, just sediment . . . and then they passed over the lip of the impact crater. Nico was saying, 'Show me wood, show me metal. Show me a piece of wood.' He kept repeating it. And then he got it.

First they saw three long pieces of wood. Robbie took a quick glance with the vehicle in the direction of the wood, then kept moving. He was getting really worried about the battery. Because of declining power, he couldn't dawdle to look at things in the way they would have liked. He just had to keep going.

And then, about a minute and a half into the dive, they came upon a long dune of mud with some debris in and around it. They looked along its length for a bit, just to confirm it was the edge of the impact crater, and then Robbie again resumed his line into the ship. Another minute and a half later, in the shadows just beyond the headlight field, they started to see the actual ship. To begin with it was just a ghostly, half-visible outline – then, suddenly, a wall of wood was before them.

Robbie took the vehicle right up to the side of the ship. The picture was so clear, they could even see her paint. Her state of preservation was incredible. They could see the seams between the planks and even the nails that fastened them.

At first Clément was a bit puzzled, as he could see two horizontal bands on the side of the ship; it wasn't until they got in close that he realized what these were. Her paintwork was in two tiers. If you look at old photos, the lower hull was black while the upper sides were white. He couldn't believe they were seeing the actual paintwork. There were no marine deposits; there was no corrosion. Everything was so fresh and clean. It was as if somebody had just scrubbed her up for this moment, her big reveal.

Following Nico's instructions, Robbie then guided the Sabertooth up and over her side and onto her deck. Because of distortion with the fisheye lens, straight lines appeared curved. There was an open hatch and he could see steps going down. He turned the vehicle a bit to the right, as he could see some broken masts and other collapsed timbers, but then he turned left again so that it was facing towards the bow, putting the vehicle on the same heading as the ship.

As they passed over the bow, Robbie, who had been really concerned about power, said the battery was extremely low. So Nico said, 'End of dive. Back to deck.'

They were so low on power, they had to pull the vehicle the full 3,000 metres back up to the *Agulhas* by its tether. They had to conserve enough battery so that once it was back on the surface, there would be sufficient in reserve to manoeuvre it into position for a pick-up off the stern.

Freddie was still on the bridge, but he knew something was going on and he needed to know what it was. He could see me and John out on the ice. We looked tiny against the berg, but because our faces were towards the ship he assumed we were on our way back. Then Nico arrived on the bridge, wondering how he could tell the captain what was happening without anyone else hearing. They crossed to the starboard side of the bridge where nobody was around. 'You mustn't tell anybody until Mensun and John are back,' said Nico, 'but we have found the *Endurance* and she's incredible.' He showed Freddie the picture. Freddie then called Captain Knowledge.

Captain Knowledge ran up to the bridge, passing Nico on the stairs. 'Is it true?' he said.

'Yes, yes, yes,' Nico replied, then took out his phone again and showed him the high-frequency view from above. Knowledge was amazed. Amazed at her condition – and that she was upright, just as I had predicted. Was this really the ship that everyone said had been crushed by the ice? She didn't look very crushed, he thought. In fact, she looked as if someone had laid her out gently on the silt and said, 'Wait here now, wait until somebody finds you.'

Up on the bridge, the two captains watched as John and I strolled back towards the ship at a leisurely pace, as if we were out on the beach. As they passed the cadets and other officers at the surrounding consoles, Freddie said to Knowledge, 'If only they knew what we know!'

By the terms of the protocols, I was the one who had to declare the wreck found, but obviously things had moved beyond that on the ship. So Knowledge went to the next stage and called the vessel's managers in Cape Town. And then he called his wife. He said, 'We have found her, but you absolutely must keep it to yourself. Nobody must know.'

John and I had no idea what had been going on while we were out on the ice. When we got back at 1645, all I was thinking about was getting out of my polar gear and warming up. My feet and hands were numb.

No sooner had I set foot on the ship than one of the cadets came up and said, 'Captain Knowledge asked me to convey his compliments and to say that your presence is required on the bridge immediately.' I was already taking off my polar gear. I asked if I had time to warm up first. The words had barely passed my lips when the tannoy crackled to life. '*Shears and Bound to the bridge immediately. Shears and Bound to the bridge immediately.*'

We raced up to the bridge. Clearly, something major had happened. We had never been summoned like this before. We suddenly became worried that maybe we had lost a submersible. We both had extremely painful memories of losing AUV 7 in 2019. A surge of excoriating retro-agony rose up inside me; I felt like a matador in that split second when he knows he is about to be gored.

But then, out of the corner of my eye, I saw something that changed everything. One of the data analysts was standing in a doorway, smirking. If we had just lost a vehicle, his face would have been ashen. And then, through the window of the ops room, I could make out Jim – and he was laughing. Suddenly I felt a contrasting rush of optimism. *Could it be? Dare I think it? Dare I?*

We came tumbling onto the bridge not more than a minute after Knowledge had put out the message. Nico and Knowledge were standing in the central part of the bridge at the port end of the main console. I just knew, now, that something huge was afoot. And that it was good, way beyond good. My blood was pumping and the distant hum of the engine room sounded more like somebody holding down the keys of a big chord on a cathedral organ.

Then came something we had joked about for years. There was Nico, thrusting the screen of his phone towards me. There was a picture on it. 'Gents,' he said, 'let me introduce you to the *Endurance*!'

Strangely, I can't really remember what I said, thought or did next. This was the lightning-rod moment I had dreamed of for 10 long years and yet, when it struck, I somehow wasn't ready for it. My memory of it splinters into a series of brief, gilded moments that don't really hang together in proper order. I remember rejoicing with John, Knowledge and Nico. We were South Africans, Brits, French and even a penguin from the Falklands: a close-knit international team who had shared a dream, and now, by this single act of discovery, we were forever joined at the hip. We shook hands, embraced and laughed and, yes, there were a few tears.

Did I regret having been out on the ice at the crucial moment? I have to say that the thought never even crossed my mind. There was simply no time, and I was so overcome with jubilation and relief. We had done it.

I remember thinking, where was Freddie? This was his moment too. But, of course, he was on watch. He was responsible for the care of the ship and we were still at the heart of the Weddell Sea pack, that most atrociously ship-hungry spot on earth that would, if it could, snuff us out like a bug. And then, suddenly, there he was, in his whites and epaulettes, walking towards me. He extended his hand and I met it with mine. He asked me how I felt now that the quest was over; I told him I could feel the breath of Shackleton himself on the back of my neck.

In the documentary of the 2019 search for the *Endurance* I am on record as saying how I felt my whole life had been narrowing to that moment. And that was true; it had. Everything that had gone before had been preparation for this day. Now the wait was over and I had reached the end, a point of perfect convergence when everything dissolved into a single, dizzying, incandescent sunburst of pure discovery.

The next thing I remember is walking past the moon pool and down the corridor to the back deck with several of the ship's crew clapping me on the back. The bosun was one. We came to the back deck, and there was Chad. We have been through quite a bit over the years and it felt completely right that he was there now. He reminded me that he had told me he could smell the *Endurance*. I reminded him that he hadn't actually said it to me *today*. He called me a miserable bastard and claimed he would have said it again if he'd seen me. We embraced and walked over to the shelter.

At that moment, the vehicle was on its way back up. There was no room in there for anything other than some awkward high-fives, hand-grasps and congratulations. Robbie reminded me that he had said 'today was the day'. I pointed out that it didn't really count, because he said it every day. We laughed, but at that moment we would have laughed at anything, we were such a happy bunch of deep-ocean explorers.

They showed me the video they had just taken. It was only four minutes long but, like everyone, I was blown away by what I saw. In my life I have had some incredible moments underwater. I have seen things that nobody ever gets to see: great works of art lying on the seabed, chests spilling with treasure, skulls gaping up at me through the sediment. I remember finding a gold coin on the *Mary Rose* and slabs of gold mixed with Ming porcelain on a wreck off East Africa. There was a Greek pot from 600 BC with two combatting soldiers on it, and as I picked it up I recognized the hand of the man who had painted it. There was the time when a friend found an intaglio on the wreck of Lord Nelson's famous ship the *Agamemnon*, which actually had the great man's name on it; there was the statuette of the rising dragon that is now on display in the British Museum; and I recall vividly the day I raised a gun from the pocket battleship *Admiral Graf Spee* that is now on display outside the Maritime Museum in Montevideo. These are moments of pure, undiluted astonishment that make your synapses crackle with surprise and wonder; moments that send you tumbling back over the decades and centuries; precious moments when you feel you have made some kind of mind-touch with people from other eras and civilizations. But nothing, I tell you, *nothing* compares with finding the *Endurance*.

As I watched the video there were several catch-your-breath moments. When we passed over the main deck, we could see a large oblong outline on the planking below. I knew immediately what it was. It was where the deckhouse had been that contained the wardroom or saloon, with the ship's galley and a pantry at its forward end. It had contained four tables, with stand-up shelves between the tables that held back-to-back rows of books. The roof of the galley and wardroom served as the ship's bridge, from where there were walkways connecting this deck with the poop deck aft.

Worsley told us that during the vessel's final death throes, when the masts were coming down, the foremast had smashed the galley and wardroom; and, sure enough, the evidence was here to see. The mast was lying to port, pointing somewhat aft. The forward side or pantry end of the building had been crushed by it. The other three walls and roof had gone. Presumably they had been salvaged for their timber. Shackleton himself tells us that deck cargo had been pushed through into the wardroom and that its starboard wall had been displaced.

As the Sabertooth passed over the fo'c'sle deck at the bow I could see that the jib-boom had gone. This was hardly a surprise, as Shackleton, Worsley and others had commented on its loss.

As an archaeologist I am supposed to exercise a certain amount of professional detachment from the object of my study, and yet what I had just seen was utterly intoxicating. The thought which consumed me the most was that 3,000 metres beneath our keel we had a sealed box containing an Aladdin's cave of polar treasures, a King Tut's tomb of wonderful things: our single greatest and most authentic trove of artefacts from the Heroic Age of Antarctic exploration.

I could have carried on rerunning that video, but then somebody appeared at the door asking if there was anybody who had not eaten. I couldn't see who it was through the pack of bodies, but he said they were about to stop serving and if we wanted dinner we had better get our arses there fast. In all the excitement several of us had forgotten about food, so we dashed off, knowing that if we didn't eat now there would be nothing until breakfast.

I sent my wife, Jo, the picture Nico had shown me. Beneath it I wrote 'Behold the *Endurance.*' And then I called her. Jo had received the picture just as she was entering the Playhouse theatre in the centre of Oxford. In the middle of the foyer, she broke down in tears. The friends she was with went in to take their seats while she went to the bar on the mezzanine floor, where she sat out the entire first act trying to get a grip on herself.

At dinner, I just sat there – all ambition spent, but happy, really happy. Chad was sitting opposite. We studied the menu. It read:

S.A. AGULHAS II

5 MARCH 2022

MENU: DINNER

Soup of the day

Mix grill: Lamb chops

Mutton sausage

Chicken drumsticks

French fries

Ice cream

I took the menu from its case as a memento of the day and Chad did the same from one of the other tables. From that point on, our conversation became more and more weird. It was almost as if we were high on undiluted joy. We just wanted to laugh and rejoice, and any old joke would do.

After we took the menus and were waiting to be served, Chad suddenly said, 'Goddammit, I gotta get me some grandchildren. What's the point of finding the *Endurance* if you don't have grandchildren to tell the story to?' So I told Chad about a little moment in Macklin's unpublished diary in which he was thinking about what he would do in old age if he survived. He saw himself sitting beside an inglenook fireplace, telling his grandchildren the story of the sinking of the *Endurance* and how they survived on the ice.

It had been a day full of surreal moments, and there was one more to come. Shawaal, who was cabin steward by day but waiter by evening, came over to take my order. 'Well done,' he said. I thanked him and ordered everything from the menu except the sausage. When I finished the main course he brought the dessert. Nothing fancy, just two globs of vanilla ice cream in the middle of a dish. As he put it before me he bent down low and, in a rather conspiratorial manner so that nobody else would hear, offered me an extra scoop.

I glanced up in surprise. In all my time on the *Agulhas*, this had never happened before. Morgan, the chief steward, was standing where he always stood at the entrance to the dining room, looking on. He gave me a wink and a thumbs-up to indicate his approval.

And so I took my reward.

POSITION: Over the *Endurance*

6 MARCH 2022

First visual inspection of the Endurance

Last night there was singing in the saloon beside my cabin, led by a guitarist from one of the helicopter crews. I did not join them but I left my door slightly ajar so I could enjoy the music. It went on until about 0200. Nobody wanted to go to bed. They just wanted to prolong the day, so when there was nothing left to say, they sang. It reminded me of the day when Shackleton's men were rescued from Elephant Island on the Chilean tug *Yelcho*. That evening, they also did not want to sleep. They just wanted to smoke, eat peaches, drink white wine and, above all, sing. And they sang their hearts out until, as Wordie put it, 'drowsiness and sheer weariness won the day'. So it was with us last night.

There were two overwhelming questions that I wanted to address alone in the relative quiet of my room. First, was the *Endurance* within my original 2019 search area? And second, if she was there, had she been found by AUV 7 before it disappeared?

The first question was particularly important to me because after we got back from the 2019 season, doubts were raised about the search design. This, by inference, impugned the quality of my research, something that had taken me many months and of which I was proud. The implied criticism, however, was well intentioned, so all I could do was grit my teeth and take it, with a smile. It will be remembered that in 2019, by the time we had beaten our way through to the zone, we had only 50 hours of search time before the *Agulhas* had to leave. That period allowed for one full dive of 42 hours plus one more brief one of about eight hours.

As soon as I got to my room, I checked the position of the wreck against my 2019 coordinates and, to my relief, found that it was comfortably within the area of the second dive. In other words, the search design was correct; my research was good, but AUV 7 did not find the wreck. Put another way, if the AUV had completed its search, we would have discovered the *Endurance* in 2019.

There was something else about yesterday's discovery that brought me a gratifying sense of fulfilment, and this was that the wreck is disposed very much as I expected it to be. We launched the search for the *Endurance* at the Royal Geographical Society in 2018 and at that meeting I made four predictions, all of which, it now turns out, were correct.

The first was that she would be upright. This was because I knew that the drag imposed on the sinking vessel by all the broken masts and rigging, which were still attached, would ensure that she would descend keel first and perhaps slightly down at the bow where she had been over-ballasted by the crew.

The second prediction was that she would be well proud of the seabed. This was a reflection of experience; I had never seen a deep-ocean shipwreck that had been significantly absorbed into the seabed. Why, I asked myself, should the Weddell Sea be any different?

Thirdly, I predicted that she would be well preserved. There was little conjecture in this, as it is well known that cold water is an excellent preservative and that wood-boring *teredo* worms and their like cannot reproduce in low Antarctic temperatures.

And the final prediction was that she would be in a semi-intact state. This was a little more complicated. In the public imagination the *Endurance* had been crushed to smithereens by the pack, but this was not supported by what I was reading in the diaries. If they had seen her demolished by the pressure they would have said so, but they didn't. From that I concluded that her primary structure would still be performing its task. The other factor which contributed to this prediction was that we had found the constructor's plans in Norway, and they demonstrated the strength of her build. The impact of deep-water wrecks with the seabed is an explosive moment, to which some wooden ships respond by opening up longitudinally or breaking in two across the hull at its point of greatest weakness. If ever there was a timber-built hull that could withstand the violence of impact, then it was the *Endurance*.

That my predictions were correct gives me a profound sense of satisfaction. Not that it matters in the greater scheme of things, but it warms the cockles of my vanity when close colleagues on the ship give me a pat on the back for getting it right. One of the best things in life is to enjoy the respect of those whom you respect, so to be complimented in this manner by Knowledge, Freddie, John, Nico, JC and Chad means a great deal to me.

So what now? We have found the *Endurance*. If we were treasure hunters or self-seeking adventurers, we would be into the mud like hogs grubbing for truffles. But this is an archaeological project; we will not be touching anything. Our mission was to find, record, educate and disseminate through publication.

It was about 2200 when we got the Sabertooth back on deck last night. We allowed a few minutes for photos and a bit of celebration, but then it was back to work. There was much to do. The ship had to be repositioned and, above all, payload systems within the Sabertooth had to be changed and prepped for go. While one team mounted 4K cameras on the front of the vehicle, another took off the panelling to remove the side-scan sonar and multibeam echo sounder in order to create space for the installation within the undercarriage of the 3D laser camera. Normally it would take about three hours to 'turn the vehicle around'; this time it took almost six. In the meantime, the midnight-to-midday team came on shift. They had slept through all the excitement.

The new team comprised Greg Morizet, Fred Soul, Maeva Onde, Fréd Bassemayousse, Kerry Taylor and Joe Leek. Most were up before they were due on deck.

The first within the new shift to learn of the discovery was Greg, a 40-year-old hydrographic surveyor and sonar specialist from Normandy. 'It was probably the most explosive moment of my life,' he later said. As soon as he heard he forgot his coffee and ran down to Level 3, where he saw JC, who showed him the first low-resolution sonar image and then the multibeam. His first thought was to wake his friends to share the moment, but he also wanted to give them a bit more time to sleep. The upcoming dive was going to be long, and extremely demanding on their concentration.

He let half an hour pass and then he woke Fred Soul. They have been friends for 15 years; they studied and boarded together and over a number of projects have, in Fred's words, 'shared the same road'. Fred was already awake when Greg knocked. His sleep had been fitful. Usually everybody is very quiet in the accommodation areas so as not to disturb the sleepers, but during the last few hours there had been raised voices, banging doors and even the sound of running feet. He was puzzled, even a bit uneasy. And now here was Greg waking him, something he never did. He opened the door. Greg was smiling.

'I think you should come,' he said. He didn't explain why, but then he said it again: 'I think you should come.' With that, he disappeared. By this point, Fred had begun to work it out for himself. Putting on his clothes, he went downstairs. Everybody he passed was jubilant, but he just wanted to get to his French friends in the shelter, where they all shook hands and embraced.

Greg, in the meantime, had gone on to wake up Maeva, a thirty-something-year-old from just outside Cherbourg. Being part of this project means more to Maeva than perhaps anybody else on board. She has read both of Shackleton's books, and Worsley's. She is obsessed with Shackleton. At Christmas 2019, just as the Falklands Maritime Heritage Trust was starting to gear up for its second campaign, Fred gave Maeva a drawing of the *Endurance*. At that point she had not yet been selected for the team. On the back Fred had written *'Je te souhaite de la trouver'* ('I would like you to find her'). Her dream was about to come true.

Maeva later told me, 'When I got to the shelter, all my French friends were already there. They showed me the scans. I burst into tears. I looked at my friends. *"On la sait, mes amis, c'est historique."* ["We know it, my friends, it's a historic moment."] And then I cried some more. I thought about how the last people to see the *Endurance* were Shackleton and his crew. And now I was part of that story. I couldn't stop crying.'

By midday, refitted and repowered, the Sabertooth was ready to go. There followed a brief trial dive to 10 metres to check buoyancy and balance. This went well, and by 1230 the vehicle was nose down and heading vertically for the bottom, 3,008 metres below. Our first priority was to secure the 3D laser-cloud data, after which we would commence the visual inspection with the 4K cameras.

The laser dive lasted until 1800. It was the most important dive of them all. It's from this dive that we will acquire the data to build the photogrammetric and 3D models of the wreck, and it is from these that I will extract the information I need to conduct a full archaeological study.

For four hours everyone working on the dive had to be completely focused, as they would be on close manual piloting the whole time. They were only going to be 2 to 2.5 metres above a wreck covered in broken masts, ropes and other clutter. Furthermore, all lights would be off; they would be operating in absolute darkness. They would only be able to see their way by the blink of the camera flashes. Just one slip, and it could all be over. Everything had to be perfect. It was, said Greg, the riskiest and most anxious dive he had ever known.

The other constant concern was the tether. This is important on any dive, but never more so than on this one, as they had to maintain just enough tension on the cable so that it always went upwards rather than looping around the wreck, where it would soon become entangled and break the fibre-optic link. At times, they were flying a circuit of the whole thing with the Sabertooth at an angle of 45 degrees. I have never known a more stressful remotely operated dive, nor one that promised greater archaeological rewards. As the camera flashed, the images showed on the screen but were quickly gone as the submersible moved on and everything returned to darkness. Greg said the images that appeared with the flashes were utterly awesome, and that none of the trio in front of the screens had ever witnessed detail and clarity like it before. By the end of the mission they had taken over 25,000 still photographs.

At the end of that dive there was no time to relax. While there was still battery power, we had to commence the 4K camera visual inspection. This was what I had been waiting for. During the previous dives it was not possible to pause operations, but on this dive we could examine anything. However, there were limits. Because of concerns over diminishing battery life we could not put the vehicle in vertical hover mode to look down hatches or through holes in the deck.

When I started keeping this record three years ago, I never knew where it would take me. The challenge before us was huge and if we did not find the *Endurance* my commentary would be of no interest to anybody. As long as there was still a chance of success, however slight, I kept the words going, but never with any sense that what I wrote would go beyond myself.

All that has now changed. Suddenly my narrative has purpose and for the first time, I feel as if I'm writing to be read. I have dreamed of writing about the first real-time visual appraisal of the wreck of the *Endurance*. It has taken almost 10 years to get here, and now we have, I want everybody to see what we saw.

I've therefore drafted a diagram of the *Endurance* (see Figure 2 overleaf) that shows the main decks, fittings and accoutrements. If you disregard the collapsed masts and rigging, you're left with four decks. The first deck (starting at the stern and moving forward) is the well deck, or afterdeck, from where the ship would have been steered. Immediately forward of that is the long, raised poop deck, sometimes called the quarterdeck. The main features of the poop deck are the mizzen mast, an engine-room skylight, two ventilators and the funnel and a small hatch. Beneath the poop are 12 cabins, six on either side, separated by the engine-room lightwell in the middle. Forward of the poop is the main deck, also referred to as the upper or weather deck. Moving towards the bow, the main features of this deck are the main mast, main hatch, deckhouse

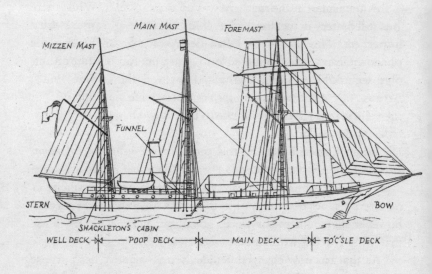

MIZZEN MAST   MAIN MAST   FOREMAST

FUNNEL

STERN   BOW

SHACKLETON'S CABIN

WELL DECK ▷◁ — POOP DECK — ▷◁ — MAIN DECK — ▷◁ — FO'C'SLE DECK

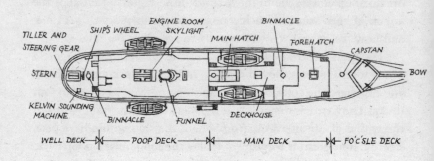

ENGINE ROOM
SKYLIGHT   BINNACLE

SHIP'S WHEEL   MAIN HATCH   FOREHATCH

TILLER AND
STEERING GEAR   CAPSTAN

STERN   BOW

KELVIN SOUNDING
MACHINE   BINNACLE   FUNNEL   DECKHOUSE

WELL DECK ▷◁ — POOP DECK — ▷◁ — MAIN DECK — ▷◁ — FO'C'SLE DECK

Figure 2. A diagram of the Endurance

(comprising wardroom, galley and pantry), foremast and forehatch. The roof of the deckhouse also functioned as the bridge deck, from where the captain and officers commanded the ship and gave orders to the helmsman in the well deck at the stern. The final deck was the raised area at the front of the vessel, where the ship's sides converged to form the bow. This was the fo'c'sle deck, or head, where the capstan was situated and where the ship's two anchors were secured when not in use.

The stern

The instant I saw the stern, I knew this would become the image of the wreck that would forever be emblazoned upon the public imagination. Just as the bow became the iconic view of the *Titanic*, so too will the stern become that of the *Endurance*. The first features to be illuminated by our lights were the nine shining, raised letters of the ship's name. There was no corrosion, so clearly they were of a non-ferrous metal, presumably bronze. Beneath the arc of the ship's name was a five-pointed star representing Polaris, the Pole Star of the northern hemisphere. It was after this star that the ship had originally been named. Shackleton decided to leave the Polaris star on the stern.

The *Endurance* has what is called a counter-stern, that is to say, its topside extends well aft of the vessel's water line, leaving the upper rudder and its post exposed to view. The rudder was always seen as the ship's Achilles heel. Shackleton and Worsley knew that if the ice got a grip on the rudder, it could spell the end. The stern is a complex part of a wooden ship where the strakes, or planks, have to curve and counter-curve beneath the tuck before bonding with the stern post assembly. Inside the stern there is a density of so-called fashion timbers and deadwoods to stiffen the whole structure, which has to withstand the slamming of the ocean. It is a part of the hull that is almost impossible to repair when at sea. And, sure enough, it wasn't 'Dame Fortune' (as one of the diarists put it) who doomed the *Endurance*; it was the rudder. The ice tore it from the post, the ends of the planks were sprung, and possibly some of the backing timbers were displaced. The water poured in and, because they could not repair the damage, and despite Chippy's best efforts to stem the flow, the vessel began to fill. If I have one criticism of the ship's designer, Ole Aanderud Larsen, it is that he did not give her a retractable rudder such as was fitted to that other famous ice ship of the period, the *Fram*.

I was surprised to see the rudder just lying there, almost innocently, in the mud beneath the counter. It told the whole story of the vessel's demise. Through the shadows I could even discern the torn surface of the stern post. On the side of the rudder I could just make out, under a fine veil of silt, what appeared to be some of the iron-strop work that held the pintles which hinged the whole device to the gudgeons on the stern post. On 27 October 1915, the day they abandoned the *Endurance*, Orde-Lees wrote: 'Crash followed crash as she vibrated to and fro. By 6 p.m. the ice had reached the upper deck; at the same time the stern post was, to all intents and purposes, rent clear out of her.' That must have been the moment when the rudder separated from the ship.

The well deck

As the Sabertooth rose up towards the taffrail or hand rail that circles the stern, we could see through the openings to the well deck within. Not only was this the smallest of the ship's decks but, quite clearly, it was also the best preserved. To starboard there were a set of steps going up to the poop deck, and to the left of the steps was a companionway going down to the accommodation area below the poop. Just inside its double doors we could see a series of pigeonholes where they kept the signal flags that could be run up to the mizzen gaff from the well. Either side of the companionway were two portholes.

The dominant feature of the well deck was, of course, the

ship's 10-spoke wheel, or helm, and her steering mechanism abaft. The ship's wheel is probably the most emblematic part of any vessel from the era of sail. As far as I could make out, it was intact.

With the exception of the cook, the cook's helper, and possibly Shackleton himself, they all took their turn at the helm. Their big fear when gripping its spoke-ends was that a chunk of ice might slam against the rudder, causing the wheel to 'kick'. To hold the ship on course, the helmsman always had to grip the wheel firmly with both hands, so any powerful contending movement of the wheel could have injurious consequences. Orde-Lees describes two such incidents. Both occurred in December 1914, a little more than a week after they entered the ice.

Of the first, he says: 'One of the members on watch today [presumably Orde-Lees himself] nearly had the exciting experience of being "thrown over the wheel" for a floe did actually strike the rudder a terrific blow and the wheel literally swung around with him holding on to it, but just as a complete summersault seemed inevitable he had the presence of mind to get his foot on the brake and check it, so preventing what might have been considerable damage to the steering gear. Sir Ernest was standing by at the time, and though he neither blamed nor complimented the member on the feat, he was good enough to congratulate him on a lucky escape.'

Two days later, on 17 December, it happened again: 'Got into very heavy pack again in the afternoon and could only get through it by repeatedly charging at it at full speed . . . Our smallest member, Mr Hussey, the meteorologist, a charming little person, was thrown completely over the wheel today, fortunately without entertaining anything worse than a shaking.'

Some enjoyed their turn at the wheel more than others. On the voyage out from England, Orde-Lees confided to his diary that he was 'very bad at steering, but it is not the sort of work to interest me – keeping one's eyes glued on a compass for two hours, and turning a heavy wheel backwards and forwards every time a line gets away from a triangular dot. One doesn't have to look at anything else. The idea of having a sharp look out for rocks, other ships and lighthouses, and steering clear of them, is quite a myth.'

Orde-Lees's incompetent helmsmanship did not go unnoticed by the crew and on Christmas Day 1914, Hurley sang a comic song, the lyrics of which made much of his 'eccentricities at the wheel'.

Wordie, whose turn at the helm was from 8 to 10 every evening, did not like the way the wheel was so low that the helmsman could not fully see ahead of the vessel. He also complained about it being 'drafty work', but then there were other occasions when he found the experience highly enjoyable. He described those moments when he had the afterdeck to himself, and how he would savour the peace as he watched the seabirds gliding back and forth across his wake.

The main joy of the day was a Kelvin sounding machine that was situated behind and to starboard of the wheel. Later, I was shown newly processed imagery that, incredibly, allowed me to read much of which was on its dial. The main words were:

KELVIN & JAMES WHITE LTD.

PATENTS

SIR W(M) THOMSON

—beneath which could be read GLASGOW. William Thomson was a physicist and engineer who was knighted by Queen Victoria in 1866 and elevated to the peerage in 1892, becoming 1st Baron Kelvin. In 1900, the firm Kelvin and James Limited was formed to produce a range of mainly marine equipment, including sounding machines.

In the old days, to aid navigation, a man would stand in the chains on the side of a vessel and swing the 'lead', a rope with a sinker of about 14 pounds at the end. The length that the line sank in a cast was the measure of your depth. During the age of steam, however, when ships were often travelling at speeds of over 10 knots, a better method was required that allowed the vessel to sound without pausing passage. In 1876 Thomson devised a sounding machine which, although it went through various modifications, was still being used in the 1960s.

It consisted of a frame containing a reel that took the sounding wire, a crank handle for rewinding, and an enamel dial (graduated to 300 fathoms) with a needle to indicate the amount of wire that had left the drum. To take a cast, the line with the sinker was paid out. As the sinker touched the water, the pointer on the dial was zeroed; the order was then given to 'let go', and the brake was released. As the line ran out a brass finger-pin, or 'feeler', was pressed gently against the sounding wire. The moment the sinker touched bottom, which rarely took longer than a minute, the line went slack and the man with the feeler called out the number of fathoms on the dial, at the same time applying the brake to prevent any further escape of wire. Because the vessel would be moving, the reading on the dial gave only an approximation of the depth. For this reason the officer then consulted a table which allowed him to match the figure on the dial with the ship's speed, so as to obtain a more accurate measure of vertical depth.

The Kelvin sounding apparatus was also much used by the expedition's marine biologist, Robbie Clark, when he was making his trawls to sample the sea water and obtain temperature readings.

During the period when the *Endurance* was icebound, the men tried to keep a patch of open water about the stern in order to protect the rudder. This provided an ideal opening through which, by using the Kelvin sounding machine, they could sample the life and geology of the seabed. The dredge would be set down on the bottom and the drift of the ship would then drag it through the sediment. Sometimes there were surprises, as on 26 January 1915, when they found a very large brittle starfish with its long, snakelike arms wrapped around the piano wire.

I had with me a large book containing all Hurley's photographs. Everybody was borrowing it. When we saw the wreck for the first time, several noticed that the topside of the stern in the old pictures looked nothing like what we had found on the seabed. This was because, while they were icebound, Chippy built a wheelhouse or hut over the afterdeck to house the scientific equipment and general gear, on the roof of which was Hussey's anemometer, which he used to gauge wind speed and direction. When they abandoned the *Endurance* Chippy took down this hut and reconstructed it at Ocean Camp, where it was used for stores. Orde-Lees describes how on 2 November 1915, a team of salvors under Wild 'contrived to collect . . . the wheel house, complete, and to bring it here on a sledge'. By then Orde-Lees had fallen out with Worsley and the others in his tent and moved into the wheelhouse, which had come to be known as the 'rabbit hutch'.

Following the Sabertooth during the inspection dive, we moved our attention to the poop or quarterdeck, within which were situated the cabins of the captain, officers and passengers. As noted earlier, its main features were the mizzen mast, the engine-room skylight, the two ventilators, the funnel and a small hatch.

The poop deck

Within the floe, the *Endurance* was heeled to starboard so that the ice overrode the hull from starboard to port, bulldozing all before it, including most of the poop deck. Also, the hull was down at the bow and raised at the stern, so the ice never fully reached the well deck nor the rear of the poop. The mizzen, or aftermast, had fallen to starboard across the poop and over what was Shackleton's cabin. The rupture across the deck passed roughly through the engine-room skylight. As for the skylight itself, although displaced and dismembered, it still survived.

As our inspection dive continued, I found myself thinking of the different characters who had once inhabited the ship. When I saw the damaged rudder I thought of Chippy and his desperate efforts to save the ship; when I saw the wheel I thought of Wordie spinning it in the film; when I saw the companionway I thought of Orde-Lees standing there, as he does in one of Hurley's photographs; and, of course, when I saw the galley, how could I not think of the cook, Charlie Green, and his assistant, the stowaway Perce Blackborow?

But when I beheld the skylight, it was Shackleton who rose within my mind's eye. Just before leaving the *Endurance* on the day they abandoned her, he wrote of how he was on the poop deck and 'looked down the engine-room skylight as I stood on the quivering deck, and saw the engines dropping sideways as the stays and bed-plates gave way. I cannot describe the impression of relentless destruction that was forced upon me as I looked down and around. The floes, with the force of millions of tons of moving ice behind them, were simply annihilating the ship.'

As for the funnel, the most emblematic part of the ship's profile, it lies in plain sight, pointing out over the starboard side. Its purpose, as the culmination of the smoke-box at the rear of the vessel's single boiler, was to project smoke and fumes as far from the deck as possible so that they would be carried safely away by the wind. Attached to the funnel was the steam whistle, and also clearly visible were the lugs to which the wire stays that held the funnel upright had been shackled.

The funnel is given a few brief mentions in the diaries. In Hurley's famous photo of Wild contemplating the ruins of the *Endurance*, taken on 8 November 1915, we can see that the funnel has a pronounced lean to starboard and that all its starboard stays appear to have been severed by the ice, while those on the port side hang limply down. Shackleton's men couldn't quite believe that when all else had gone, the funnel, one of the weakest parts of the ship, was still standing. On 14 November, Orde-Lees wrote: 'Nothing but the funnel is now visible from our camp . . . Her end is near. Soon she will be gone.'

Two days later, he added: '[T]he funnel leans over to starboard and will soon fall, and all three masts have already gone. It is an unpleasant sight, depressing in the extreme. To think that this is all that is left of what we had until so recently reckoned upon being such a stout little craft, and in which so many happy days had been spent by all of us.'

But the funnel didn't fall – it was still standing when she sank on 21 November. In fact, as we know, it was a movement of the funnel that caught Shackleton's attention and alerted him to the fact that she was about to make her final plunge.

The cabins below the poop deck

Within the poop, there were six cabins on either side. Outside each row of cabins was a passageway, the inside wall of which was the bulkhead of the lightwell below the engine-room skylight. The passageways connected with the well deck companionway at rear, and forrard with doors that gave out onto the main deck. Orde-Lees found these passageways uncomfortably tight: 'It is extremely difficult to walk about in the ship, especially in the passages, which are narrow and therefore not wide enough to enable one to lean over sufficiently to maintain one's balance and so one goes along "bump . . . Bump . . . bump" against the wall all the time.'

After 10 March 1915, when winter was upon them and the noonday temperature within the cabins was down to −6°C, the people staying in those rooms moved to the main hold deck ('the Ritz') or the wardroom ('the Stables'), where there was better heat and insulation. Seven months later, with the return of spring, they all returned to their old quarters beneath the poop. During the winter months, the only person sleeping in this part of the ship was Shackleton.

Along with the wardroom, the scientists' cabin was very much the heart and soul of the ship. It was where they frequently gathered for sing-songs in the evening. According to Wordie, it was named 'the Rookery' on 4 November 1914, the day before they reached South Georgia, and the singing that went on there was called 'caws': 'For the evening it was called the "Rookery" and it may well be that the name will stick to the cabin . . . [we] smoked until we could hardly see in the small cabin. The evening was a great success; no shanties were sung, but mainly songs from the *Gaudeamus* and the Scottish students' song book.'

Within the accommodation area, the two main quarters were located at the rear, where they backed onto the well deck. They were separated by the passageway leading from the well deck companionway. The companionway was not situated over the ship's centre line, but rather was slightly to starboard. This meant that the cabin to port was larger than that to starboard. The reason for this was that the ship's designers had intended the portside cabin to be the saloon, or so-called smoking room, where de Gerlache had envisioned his wealthy passengers gathering with their cigars to read and play cards. Shackleton, however, was desperate for bunk space and because he could not sacrifice any of the vessel's stowage area, the only solution was to convert the saloon into an eight-berth cabin for the scientists.

The cabin across from the Rookery, on the starboard side of the ship, was the captain's. If anybody had asked what, for me, were the three highlights of the archaeological dive, I would without hesitation have named them as the Kelvin sounding machine, the three holes in the main deck (to which I will return later) and, above all, my first sight of portholes that I knew immediately were those of the captain's cabin. It was a moment that set my nerves crackling because this was the *sanctum sanctorum*, the sanctuary of sanctuaries: Shackleton's cabin.

On the voyage out from England it had been Worsley's cabin, but Worsley's captaincy had not been a great success. At times, when 'splicing the main brace', he had allowed the alcohol to flow and some of the men got 'mad drunk'. One man was only prevented from jumping overboard by Tom Crean grabbing him by the leg and, while in Madeira, four of the men became drunk, damaged a bar and were taken to jail, where one was flogged. Another black mark against Worsley was when they arrived in Montevideo burning their spare spars, one of the science huts and 'all the available wood' because they were out of fuel from not having coaled at Tenerife. When Shackleton arrived by steamer at Buenos Aires on 17 October he was not at all happy with the way things had gone and took over the captaincy of the *Endurance*, although Worsley maintained his status and was still called 'Skipper'. At that point, Shackleton moved into the captain's cabin.

We know from Orde-Lees that the cabin was 'about 9 foot by 5 foot or rather less'. We also know from a photograph by Hurley that it had two portholes to starboard, between which Shackleton

had affixed a framed copy of Kipling's poem 'If'. It was also heated
with a stove and, according to Orde-Lees, Marston had been 'busy
doing up Sir Ernest's cabin, the original captain's cabin, and is doing
it remarkably well': 'He has upholstered the settee most tastefully,
using the back of an old green plush curtain, and has put up some
excellent book shelves. The stove is in place and he is making a
stout padded door. I really think that Sir Ernest will not be so
uncomfortable after all, but I am afraid that his lonely existence
will give him time to brood over this great misfortune that has
befallen the ship – failure to reach our proposed base or indeed,
any land.'

Although the very aftermost part of the poop structure, that
is to say, the end nearest to the well deck, is roughly intact,
everything forward, particularly on the starboard side, has been
completely disrupted if not erased. There are several reasons for
this. First, the collapse of the mizzen or after mast on 18 October
1915, would have imposed a levering action on the surrounding
deck. Wordie, for instance, recorded on 1 November that some of
the masts had been cut away and 'the mizzen [was] tearing up the
rookery'. Four days later he wrote that 'the Rookery was quite
upside down, the mizzen having heeled over there', but nonetheless
he was still able to get into his old cabin and recover a few books.

At the same time, with the hull heeling to starboard, the ice
was overwhelming its starboard side and crushing the poop and its
under-deck accommodation to port. Again, this is attested in the
diaries. On the day they abandoned the *Endurance* (27 October 1915),
Worsley wrote: 'The ship was not abandoned one hour too soon,
for shortly after we had camped on the floe we could hear the
crushing and smashing up of her beams and timbers and subsequent
examination of her showed that only six cabins aboard [those along
the portside] had not been pierced by floes and blocks of ice.'

In other words, the upper structure on the starboard side was
being bulldozed across the main deck to port. On 29 October, two
days after they abandoned ship, Macklin went back to his cabin to
try to save one of his diaries and a Bible his mother had given him

as a boy. He wrote: 'The starboard cabin ours (McIlroy's and my own) are burst in [from starboard] and pressed against the engine room bulkhead.' The same day, Hurley wrote: 'The floes, like a mighty vice, have crushed her laterally. The starboard side has been crushed in, and all the cabins along it have been closed up as efficiently as a folding Kodak. Their broken and splintered walls being forced into the alley ways, which are blocked with an indescribable chaos of debris and ice.'

This was a matter of some sadness to me, as I knew those starboard cabins had contained most of the personal belongings of their occupants. Some spoke of the loss of their suitcases, their family photos, the books they had been reading, as well as what Macklin called their 'many treasures'. Macklin made three attempts to enter his cabin, but did not succeed. In my own account of our 2019 expedition, I wrote that the one thing I would most like to find would be that lost first volume of Macklin's diaries; but today, when I saw the remains of the poop deck starboard cabins, I realized that this would never happen because they had been completely expunged by the ice. One could see, from their footings on the deck, exactly where the first three cabins had been.

The main deck

Leaving the under-poop accommodation area, we come out on the main deck at the waist of the ship, where the first feature is the main mast. Like the mizzen, this carried only fore and aft sail. According to Hurley, who filmed it, this mast came down on 28 October 1915.

Forward of the mast is the open hatch that we first saw yesterday. There were two cargo hatches on the main deck, of which this was the larger, the so-called main hatch. It also acted as an accessway to the 'tween deck, and as the Sabertooth passed over it we could see steps going down. Beside the main hold Chippy had built pens for two pigs they had acquired at South Georgia; later these were used to hold penguins awaiting slaughter.

Immediately forward of the main hold was the deckhouse with its three parts: the wardroom proper or saloon and, towards the bow, the galley and pantry. Beside the pantry was a set of steps going down to the 'tween deck below. Connecting the galley with the saloon was a serving hatch. The wardroom, according to one diarist, measured 12 by 18 feet and contained four tables, each of which sat four people. As still happens on working ships, Shackleton's men tended always to sit at the same place. On the voyage out from England, the rather haughty Orde-Lees chose to sit with Alf Cheetham and Chippy McNish because he thought that 'a little unrefined company would be a good thing' for him; but he added, 'I must say that McNish is a tough proposition. First he sucks his teeth loudly, and then he produces a match, carefully sharpens it, and proceeds to perform various dental operations. Occasionally he expectorates through the window, and at scooping up peas with a knife, he is a perfect juggler . . . But I feel sure I get on his nerves as much as he does on mine.'

At another point in his diary, Orde-Lees expressed his disgust at the way Cheetham and McNish always licked the necks of the sauce bottles so that their dribbles did not run down the sides on to the tablecloth.

Because Shackleton had converted the original saloon and smoking room into a six-berth cabin, the wardroom became the main place where the men would gather to enjoy each other's company and relax. On 21 November 1914, while at South Georgia, Wordie wrote: 'This evening, as usual on Saturdays, we have had the toast of Sweethearts and Wives in the wardroom, followed by a few songs.' 'Sweethearts and wives' is a traditional Saturday night naval toast to which somebody nearly always replies, 'May the two never meet.' With Shackleton's men, the toast often became the excuse for some musical fun, with Hussey leading on the banjo. Invariably there were occasions when spirits got a little out of hand. On 31 January 1915, Wordie wrote: 'Yesterday evening Sweethearts and Wives was extended into a proper sing-song, and carried on till 10; it might have gone further had there not been a small tendency to rowdi-ness'. Wordie was always very guarded in his choice of words, so this must have been a wild party.

For warmth there was a small bogey-type stove in the ward-room. This was one of four on the ship: the others were in Shackleton's cabin, the fo'c'sle for the crew, and the one which had been intended for use in the hut ashore but which ended up in the Ritz and then, later, at one end of the cabin area beneath the poop. During the summer the wardroom stove was little used but by the first week of February it was constantly lit.

Ice quarried from the floes was kept in a basket just outside the wardroom door, and there was an ice-melter – a huge iron oven – on top of the stove. Anybody taking water from the melter had to replace it with a like amount of ice. As might be imagined, the stove was also much used to dry wet articles of clothing.

When winter began to bite and the team vacated their cabins within the poop deck, most moved to the 'tween decks, but some went to the wardroom. Chippy built narrow cubicles there for Wild, Worsley, Crean and Marston. Because of their resemblance to horse-boxes, the wardroom then became known as the Stables.

On the bridge was a skylight to the wardroom below and a binnacle housing a compass. I knew we would not find the latter, as it had been taken ashore and erected on the lookout station at Ocean Camp. There were gangways linking the bridge to the poop deck, but while at Buenos Aires this area had been roofed over so as to create a continuous promenade deck that could house the dogs, cottage-like, in rows of kennels down either side. There was one kennel per dog, and each dog was chained so that he could not quite fight with his neighbour. As there were 70 dogs in all, the howling and faeces were considerable. At South Georgia, the deck was completed with proper planking corked with oakum and tar, thus allowing people to pass from their accommodation in the poop to the wardroom without getting wet from spray. A hatch on the new deck was created over the main hatch below, so as to allow for the movement of cargo in and out of the hold.

Apart from its forward bulkhead, the entire deckhouse has been erased, but the footings of the walls can still be seen clearly on the deck. One thing that surprised me was the number of cups and dishes that were lying around the wardroom area. Had they been bounced out of the galley when the ship struck bottom, or had the ship been raining crockery as it sank and, like leaves on the wind, a few had landed on the main deck? We know from the diarists that most of their tableware was metal. One diarist spoke of their tin plates. Another called them aluminium. On the voyage out, one commented disapprovingly on how 'we have nothing but enamel plates and cups'. They had set out from England with glasses, 'but we soon broke [them] all and since Tenerife we have had all our drinks in enamel iron cups'.

From asides in the diaries, it is possible to construct a picture of what caused the destruction of the wardroom. It began, of course, with the ice as it encroached further and further across the deck towards port. In *South*, there is a good description of the state of the wardroom on the day after they abandoned ship:

> The wardroom was three-quarters full of ice. The starboard side of the wardroom had come away. The motor-engine forward had been driven through the galley. Petrol-cases that had been stacked on the foredeck had been driven by the floe through the wall into the wardroom and had carried before them a large picture. Curiously enough, the glass of this picture had not been cracked, whereas in the immediate neighbour-hood I saw heavy iron davits that had been twisted and bent like the iron-work of a wrecked train. The ship was being crushed remorselessly.

I leave the last word to Wordie, who, on 6 November, wrote that the wardroom was 'nothing but wreckage'.

We were curious about the whereabouts of all the broken planks from the deckhouse. If some of the crockery survived, why not a bit more of the wood? Some of it would, of course, have floated off as the ship went down, but I suspect that most of it was salvaged. After all, the deckhouse walls were of light planking that would have been eminently usable at Ocean Camp. As I studied the main deck area through the eyes of the Sabertooth, I could see clear archaeological evidence for the salvage: many timbers had been cut away, and in places the saw marks were as fresh as if they had been made yesterday. The bulwarks had gone, and in some places it was plain to see that they had been sawn through rather than snapped by the ice. Even without the evidence of my eyes, there is much in the diaries regarding the salvage of the timbers. They started collecting and sawing away elements of her structure on the day they abandoned her. The following day, Wordie commented upon how the 'cutting down continues'.

One can assume that the first pieces of salvaged wood were for the sledges and other things that were needed for the great march towards the islands of the Antarctic Peninsula, which began on 30 October but ended two days later when they set up Ocean Camp only a mile and a half northwest of the *Endurance*. It was then that the serious salvage of her timbers began. On 1 November, Worsley mentioned bringing 'a quantity of lumber from the wreck'. They obviously thought this was a good thing, and the following day Worsley noted in his diary that they had brought 'as much lumber as they can cut and carry' back to the camp.

As well as being burned in their stove, the wood was also used in construction. In *South* Shackleton describes how 'planking from the deck [was] lashed across some spars stuck upright into the snow' for the lookout and presumably the galley. Orde-Lees writes of having 'laid down joists for the planking of the floor in the rabbit hutch'. Wordie speaks of laying a wooden floor and indeed, twice in his diaries Shackleton mentions the wooden floors of their tents. On 11 November, Orde-Lees writes, 'at first we had merely a very thin ground sheet between us and the snow but when so much wood came up from the wreck we constructed a wooden floor and are now able to sleep on a dry floor in considerable comfort'.

Immediately forward of the deckhouse was the collapsed main mast and to the front of that was the small hatch of the forehold, which, like the main hatch, was found open. Its cover was beside it. At the sight of the forehatch, two stories sprang to mind. The first occurred when somebody fell through it at South Georgia. As usual for anecdotes like this, it came from the pen of Orde-Lees: 'Still snowing slightly. I went down to the store room and arranged it according to my own ideas. Whilst down there one unsuspecting member trod on the lid of the hatch which wasn't there; he fell down and as I was underneath I broke his fall and he nearly broke my neck. We were both equally surprised, and I was really rather injured.'

The second account comes from Macklin and tells of the terror he experienced when he had to climb down through the forehatch on the day they abandoned ship:

I was called to the ship by Wild, but had scarcely got there when the place selected for the camp started breaking up, and we had hurriedly to harness up and move to a floe further forward on the starboard side. I returned with Wild to the ship to get up some timber from the forehold. The steps made by Chippy had carried away, the uprights were bent, and the cross beams going. We managed to get down, however. The noise inside the hull was terrific, the timbers banging away in a most unpleasant manner. The planks which Wild wanted were in the side pockets, black dark, and to enter them it was necessary to clamber through a small opening in a bulging bulkhead. I hated the thought of going in but Wild entered without hesitation, telling me to wait outside while he passed out the planks. I was glad when the job, which had taken some time, was over and we could get on deck and off the ship, for I do not think I have ever had such a horrible sickening sensation of fear as I had whilst in the hold of that breaking ship.

The fo'c'sle deck

To end the archaeological inspection dive, we came to the raised deck at the bow, also known as the fo'c'sle deck or head. This deck was accessed by two sets of wooden steps. The steps themselves have gone, although there is reason to believe they may be situated just under the edge of the deck. The deck itself has become detached from the vessel's sides and has been displaced aft as well as to port. Presumably this reflects the action of the ice that entirely overrode the bow or, maybe, a recoil reaction when the bow struck the seabed.

POSITION: Over the *Endurance*

'Lash up and stow'

In Frank Wild's words, it is time to 'lash up and stow'. Our odyssey isn't quite over, but operations are. We held our last dive this morning.

At 0700 I went into the shelter to review the footage as it was playing on the screens. I came out at 1100 with a bundle of notes and a head that was swimming. I was one blissed-out archaeologist. After four hours in that lightless, sealed container, the Antarctic noonday sun left me dazzled.

I was surprised by the number of people who had gathered on the deck to await the return of our heroic Sabertooth from its last dive. All were in their polar clothing, hard hats and steel-capped boots. There was an end-of-term atmosphere to it all.

We had found the *Endurance* on 5 March, exactly 100 years to the day since Shackleton was buried on South Georgia. Many of life's coincidences just scream out to be ignored, but this was not one of them. It was John who first spotted the strange concurrence, but within an hour it had swept the ship and everybody, from Captain Knowledge down, wanted to go to South Georgia to pay their respects to the Boss. It was, after all, but a moderate deviation from our course back to South Africa. A request was put to head office in London and today Donald Lamont, chair of the Trust, contacted us to say we had permission. His only proviso was that we have the ship back in Cape Town by the end of the charter. It was tight, but it was doable.

Everybody on board was gobsmacked by the timing of the discovery, but none more so than Fréd Bassemayousse. Fréd is one of life's pilgrims, somebody with a mystic streak and the soul of a poet whose passions are art and the sea. He was profoundly moved by the occurrence and began to wonder if there was not something more to it, something from beyond the realm of everyday reality. A sort of Shackleton 'fourth man' moment. The more he thought about it, the more he fixated upon the notion that maybe, just maybe, Shackleton had been interred at the moment we found the *Endurance* – at 1604 hours. He came to me and asked if I knew the actual time of Shackleton's burial. I didn't. I consulted what little reference material I had with me, but when this failed I thought I would go to my friend Stephen Scott-Fawcett, one of the project insiders who is host to the Sir Ernest H. Shackleton Appreciation Society, the largest community of its kind in the world. There is nobody more knowledgeable on Shackleton than Stephen.

Sure enough, within an hour, he came back to me with an answer. The pall-bearers at Shackleton's funeral had all been Shetland Islanders from the ship that had carried Shackleton's body from Montevideo back to South Georgia, where he had died on 5 January 1922. Hussey, who escorted the body, wrote:

We arrived in the whaler Woodville with the body of Sir Ernest Shackleton at South Georgia, the gate of the Antarctic, on February 26th. We took the body ashore on March 1st. The coffin was taken to the Lutheran church with floral tributes. At 3 p.m., on March 5th, the Manager of the Whaling Stations on the island, and about 100 men assembled at the church. Mr Edward Binnie, resident magistrate of South Georgia, and the customs and other officials representing the Falkland Islands Government were also present.*

I called Stephen. His view was that if the service was at 3 p.m., then surely it would have started a few minutes late. The service itself would have lasted about half an hour, the procession to the little Whalers' Cemetery would have taken another 20 minutes, a

---

* *The Times*, 4 May 1922.

few words would have been said and the interment itself would have taken place a few minutes later. They would have been burying him, he concluded, at 4 p.m. or just a little after. In other words, the same time, or very close to the same time, that we found his ship.

During the afternoon, while the crew prepared *Agulhas* for passage, we all went down onto the ice to celebrate. Tents were erected and food was served. John and I kicked my football about until it was borrowed by the crew, who wanted a proper match, so then together we wandered about shaking hands and thanking people. Freddie produced a new sign that featured everybody's signature together with the now-known coordinates for the *Endurance*. It was erected on a pole in the ice beside the ship and immediately became the focal point for photos. I gathered ice from the floe which, in my cabin, I allowed to melt so that I could bottle it for return to England, where I will refreeze it into ice cubes for consumption on special family occasions.

By mid-evening we were ready for departure. The only item of equipment not secured for sea was one of the helicopters, which will be used to recce the pack for the most suitable leads to get us out of this frozen maze. Shortly before 2200 hours, cogs deep within our great ship gathered, engaged and ground as we girded for action. I went out on the side deck for a last look. As usual, there was a small group of quietly disposed, regal-looking penguins by the edge of the pool at our stern. For as long as I could take the cold I just stood there, letting my thoughts circulate. Three days have passed since we found the *Endurance*, but in my head I am still at this intersection where dreams, fears and moonbeams meet reality. I shall miss this place. Everything down here is so full-volume and in-your-face, the weather above, the sea below and the jostling ice about us.

As the ship shook off her lethargy, the penguins stiffened and one gave a long, not-so-dulcet honk. The emperor's last call. One by one the penguins sank to their bellies and tobogganed off, imparting the prints of their feet upon the receiving snow. How different to three years ago, when we lost AUV 7: the pack was then tightening about us and there was real apprehension about getting out. There was also tension on the bridge, and an awful sense of failure pervaded the ship from stem to stern. I remember feeling like a beaten boxer, the contender that never made it. How different to now! Now I feel like one of those chesty puffins with a beak full of sand eels.

I walked to the bridge and entered by the wing door. Freddie was there. He smiled. 'We have 60 miles of pack to get through. Time to get out of here.'

Spit-spot, I thought.

If I were asked to choose what was, for me, the most uber-awesome moment of the archaeological inspection dive, I would without any hesitation say it was when I found myself staring down upon three rough-cut holes through the main deck between the portside bulwarks and the wardroom end of the deckhouse. When these holes were first spotted the day before yesterday, they were dismissed as ice damage; but, as I gently corrected, those holes were so much more than just holes. Within the long *Endurance* saga, those three holes represent a turning point in their fortunes. Three tons of food passed through those holes. Those holes saved their lives. Before those three holes, they were almost certainly going to die; after those three holes there was a chance they might not. To me, those holes and what it took to cut them epitomize the stick-at-it, spit-in-your-eye, raised-fist determination of those 28 marooned men to survive.

I found it curious to think that, despite their significance, we in the shelter were actually seeing something that Shackleton did not. He had visited the ship on 3 November when they were cutting away the upper works to reach the target area on the main deck, but during the next two days when the holes were open, he stayed at Ocean Camp. The following day there was a blizzard and all organized salvage came to an end. So when he dictated his memoirs to his amanuensis, Edward Saunders, in New Zealand, he was

mainly depending upon the diary of Orde-Lees – who also did not
see the holes, because he was the store-keeper who took charge of
all the salvaged food as it was coming into camp on the sledges.
The irony is not lost on me that, in what follows, I have an advantage over Shackleton as I have seen all the diaries, including the
ones he did not, and so perhaps have a more thorough grasp than
he did of what actually happened on the *Endurance* during that
two-day period which was so crucial to their survival.

It will be remembered that they abandoned the *Endurance* on
27 October. For three nights they pitched tents on the ice beside
the ship. This they called 'Dump Camp', but then, on 30 October
1915, they shot the cat and set out on 'the Great Journey' for the
islands of the Antarctic Peninsula, hauling the *James Caird* and one
of the cutters with them. Two days later, Shackleton realized this
was not going to work and, only a mile and a half to the northwest
of the *Endurance*, they established a base which they called Ocean
Camp. Their intention, wrote Shackleton, 'was to drift northward
to the edge of the pack and then, when the ice was loose enough,
to take to the boats and row to the nearest land'.

Now that they were going to have to sit for several months and
do nothing but wait, concerns began to grow about what they had
to eat, or rather, what they did not have to eat. On the day they
gave up the march, a party went back to the ship to recover material and, above all, food. That evening James Wordie wrote: 'In the
afternoon some men went back to the ship to try and get food
there – little success.' The following day they went back again. That
evening the expedition physicist, Jimmy James, did the maths and
was not impressed; he wrote: 'Retrieved a certain amount of food
stuff, but not very much.'

The same day, Wordie, one of the more far-sighted and
thoughtful members of the team, recorded that 'a strict tally of
provisions was made today – we have not much to come and go
on. The Boss at last has realized that we may not get out of the
pack for several months – perhaps six.' In his diary for the same
day, Shackleton wrote: 'Have food for 180 days at 1 lb per day.' It

was becoming clear to them that, if they did not reach the edge of the pack before winter, when the big freeze set in and the seals and penguins became scarce, then, slowly and one by one, they would very likely perish. There were mutterings about Shackleton's failure to save more of their supplies when he had the chance.

Two days after they abandoned ship, the ever obedient and rather servile storekeeper, Thomas Orde-Lees, wrote: 'When the water had begun to gain on the pumps, we had suggested to Sir Ernest once or twice that we should get up some of the flour, sugar and powdered milk of which we had a good supply, but he considered that it might be dangerous to have us working down in the hold. It was rather a pity that we were unable to get some of the stores up on deck at least, for now the whole of them are submerged and it will be next to impossible to get them out. The water is up to the upper wardroom tables.'

Clearly their situation was grim, but then, on the day after they gave up the march, the astute and ever pragmatic Hurley had an idea: 'The teams ply to and from the wreck, bringing into camp loads of wood and canvas, though very little food. I examined the condition of the ship, and suggested that an attempt should be made to cut a way through the deck into the Billabong in the Ritz, where a large quantity of cases of foodstuffs are stored. This will be tried tomorrow . . .'

The problem, however, was not so much that they had to penetrate the deck, but rather that the deck was submerged under 2 feet and 9 inches of water and ice (as measured by Chippy).

The next day, 3 November, a large party under Wild commenced clearing a bank of crushed ice from the port midships area, and Chippy began removing a section of the promenade deck and other upper timber works so that they could access the main deck above the Billabong. Work, however, was hampered by 'a 2 foot layer of mushy ice covering the point of attack', so it was not until the next day that they were able to start cutting through the deck.

At this point Orde-Lees wrote: 'The carpenter went off early and directed operations with so much success that he succeeded in cutting through the 3-inch deck, now 3 feet under water, and making so large a hole that many cases floated up. Others were subsequently raised by means of a boat hook. The cutting of the hole must have been terribly hard work, for the ship is very stoutly built and the deck is made of three such planking, and the most advantageous position was 2½ foot underwater.'

Hurley described their method: 'a spar [was] rigged with a heavy ice chisel to be operated as a reciprocating drill'. With his usual eye for detail, Orde-Lees gave a more fulsome explanation:

The method employed was chiselling with a large 3-inch ice chisel, sharpened up for the purpose, rigged to pile-driving tackle, and hauled up and down in the manner of a pile-driver. As soon as a large enough slip had been hacked out, a saw was inserted and gradually, by chiselling and sawing, a hole nearly 3 feet square was made about two feet from the ship's side and close to the wardroom door; that is, immediately over the corner of the hold where most of the provisions were stored.

Later on a second hole was pierced more forward, and a 'fish tackle' was fastened on to the internal woodwork between the two holes. By hoisting the 'fish tackle' the whole of the remaining wood was rent away, and the work of extracting the submerged cases proceeded. At times the men were working with their arms in ice-cold water up to the shoulder for half an hour at a shift. It is really wonderful what has been accomplished by dint of dogged perseverance, skill and toil. No less than 105 cases of provisions have been brought to the camp today, representing some two tons of provisions. What this means to us in our present situation words fail to express.

Shackleton called it the 'Fine and fortunate day.' Brief as ever, but clearly pleased, Chippy wrote, 'I have cut 2 holes through the deck – [and] we got three tons of stores out today. I am cutting another hole tomorrow.' Equally succinct was James: 'Good

progress made in breaking into the Ritz.' Others were more expan-
sive; on this day (4 November) Wordie, who was not himself at the
ship, wrote: 'The great event today was the getting of stores from
the Billabong. The first loads to reach us consisted of walnuts, curry
powder and figs – not very sustaining; but better came later – rice,
suet, hams, sugar in plenty and last of all, what we most wanted –
flour. There is much else besides; in fact the whole outlook at the
camp is now changed; instead of a meagre ration, we should have
full whacks now. Seal fried in its own blubber made a splendid
lunch today.'

The same day, Hurley described the happy mood within the
salvage team:

> The salvage party broke through the deck shortly after 11 a.m.
> The opening was followed by an outrush of walnuts, onions
> and numerous small buoyant articles. By fishing with boat
> hooks, case after case was directed to the opening from which
> they emerged buoyantly to the surface. The scene was highly
> amusing, and reminded me of the juvenile game of 'fish pond'.
> If one of the fishers brought to light a case of high food value,
> a great cheer arose. I was just in time to see a keg of soda
> carbonate greeted with groans. So, in proportion to the rela-
> tive values of the salved cases, so was their appearance greeted
> with suitable exclamations. The party worked at high pressure
> all day, taking advantage of the restful state of the ice. By
> evening, practically all the cases were fished from the Billabong,
> making aggregate of three tons of stores.
>
> All the flour was retrieved, as well as a large percentage
> of the sugar – the two commodities we were most in need
> of. The teams were busy transporting the ice-covered cases to
> the camp all day; each team averaging five loads and loaded
> to 120 lbs per dog.

Shackleton completed his diary entry for that day in his usual
telegrammatic manner: 'Hoosh in [hood of] binnacle as stove.

23 Psalm. Fine weather.' His fleeting reference to the 23 Psalm is interesting. Clearly, its words resonated. Shackleton could and did recite it from memory. His diary was small and he had filled his page. Had there been space, he would probably have written it out in full:

> The Lord is my shepherd, I lack nothing.
> He makes me lie down in green pastures,
> he leads me beside quiet waters,
> he refreshes my soul.
> He guides me along the right paths
> for his name's sake.
> Even though I walk
> through the darkest valley,
> I will fear no evil,
> for you are with me;
> your rod and your staff,
> they comfort me.
> You prepare a table before me
> in the presence of my enemies.
> You anoint my head with oil;
> my cup overflows.
> Surely your goodness and love will follow me
> all the days of my life,
> and I will dwell in the house of the Lord
> forever.

The following day, 5 November, Chippy cut the third and last hole and, once more, the stores came bobbing up or were coaxed to the surface with boat hooks.

Shackleton wrote: 'Fishing in Billabong. 38 cases of flour. Jam. Enough for 6 months food. Last day of work on ship. Overcast bad light all day.'

By evening there was little or nothing left to salvage. Wordie summarized the situation: 'The salvage work at the ship is practically finished now and it has been well worth the trouble – men working up to their knees in water retrieving stores from the Billabong with boat hooks. Today we got 30–40 cases of flour and some cases of jam; we are now well supplied for several months . . . full rations today; not so tired and hungry as yesterday.'

The next day, 6 November, Shackleton assessed the situation. Worsley recorded: 'Ernest holds a Council of War with Wild, Hurley and myself. He decides that having 6 months ½ rations outside the full amount of sledging rations that we can carry on the sledges and the seals and penguins we hope to get, we shall in the mean-time go on full rations.'

With everybody on full rations, there was across the diarists a general feeling of well-being.

It was not often that Worsley agreed with Orde-Lees on anything, but on 7 November, he wrote, 'Everyone is happy and contented and hopeful,' and later, 'I do not worry about those dangers which will probably be very great but live comfortably and happy in the present, and I can truly say that at present I am enjoying myself far more than I would in civilization.' He afterwards added: 'We are all very well fed and happy.' Wordie agreed: 'Things look more promising . . . camp seems a cheery enough place now. Hussey plays his banjo almost every night after hoosh, and generally draws forth some singers . . .'

But they had been lucky, for as Orde-Lees noted, 'It is an act of Providence that the provisions happened to be all stored on the port side, which is more uppermost.' There was another cache of stores within the ship, but it was on the starboard side and, as Chippy pointed out, was under 12 feet of water. And so, still in boxes, there it remains to this day.

POSITION: Over the *Endurance*

8 MARCH 2022

Out of the pack

Tomorrow we gift *Endurance* to the world. Today, we have been working with our London office on an embargoed press statement that will be released by the Falkland Islands Maritime Heritage Trust at 0700 GMT. From then and forever, I know that incredible view of the stern will be iconic.

Despite our euphoria, there is just a dab of sadness in the team. I myself feel it. For three days this precious wreck has been ours alone, but tomorrow it will belong to the world.

There was some stiff resistance from the pack during the night along with several hours of wind and heavy snow, but by morning we were out of hostile territory and into an area of fairly loose floes that should take us through to the ice edge. We have had a lot of gloomy weather over the past couple of weeks but this morning the clouds peeled back and great stabs of light burst through to illuminate the ice fields about us. Suddenly everything became bejewelled. The blizzard had smothered the floes in baby-fresh, untainted snow flecked only by the odd penguin or seal. This was Antarctic sorcery at its chimerical best. And if that wasn't enough, we had a bewitching glimpse of a huge whale, and the birds were back. As we breasted our way through one of the last ice fields, I went down to the stern to watch the petrels and other clunky-beaked gliders as they swooped in low over the riven ice in search of treats.

But in some respects this has also been a tough old day for me. We have had the punch-the-air and clap-each-other-on-the-back moment, which was followed by the long stare, but now, with diving operations over, it is time for the big think. I have to make sense of it all. And I must begin the process straight away, because this evening I had to give the first exposition of our findings to those on board.

I am not a natural speaker, and I felt tense at the beginning of the talk. My thoughts would not elide, I stumbled over words and I laughed a touch too generously at my own jokes. But then somebody interrupted with a question, which forced me to look the audience in the eye. They were all friends with whom I had shared something wonderful. The nerves lifted and I began, almost, to enjoy myself.

I had planned to go to bed early, as Dan Snow and I had to be on air with the BBC at 6 a.m. And after that I knew, for Dan, John and me, it would be just one interview after another. As my head touched the pillow, my phone pinged with a message. It was from Donald Lamont, requesting a statement for the Trust's website. I got up and went back into my day room. I was tired and clean out of original thoughts, but after a few minutes' consideration, I thumped out the following:

Ladies and Gentlemen,

I don't know how else to say this, so I am going to come straight to the point.

We have found the wreck of the *Endurance*!

The discovery was made at 1604 hrs GMT on 5 March 2022, in 3,008 metres of water, just over 4 nautical miles southward of Frank Worsley's famous coordinates for the sinking.

She is upright, well proud of the seabed and in an excellent state of preservation. You can even see her paintwork and count the fastenings.

Most remarkable of all is her name – E N D U R A N C E – which arcs across her stern with perfect clarity. And below is the 5-pointed Polaris star.

And in the well deck at the stern is the ship's wheel in perfect condition.

In a long career of surveying and excavating historic shipwrecks, I have never seen one as bold and beautiful as this.

The search for the *Endurance* has been ten years in the making. It was one of the most ambitious archaeological undertakings ever. It was also a huge international team effort that demonstrates what can be achieved when people work together.

Shackleton, we like to think, would have been proud of us.

Mensun Bound
(Director of Exploration)

As I again turned in, I looked out through my window and was a little surprised by the complete absence of ice. And so we are out of the pack and heading for South Georgia.

NOON POSITION: 68° 38.430' S, 050° 56.526' W

9 MARCH 2022

News of the discovery is released

Today was the day when the news broke. At precisely 0700 hours Zulu, or GMT, I hit send on several pre-prepared messages.

To my friends I wrote:

You can shout it from the rooftops. We have found the wreck of the *Endurance*.

To my boys I wrote (drawing from a poem I used to read to them when little):

Oh frabjous day! Callooh! Callay!
We have found the wreck of the *Endurance*.

To my wife, Jo (who, since we were students, has been a diver with me on wrecks all over the world), I sent something a bit different. Some lines from Browning which she already knew and which were favourites of Shackleton:

Are there not, dear Michael,
Two points in the adventure of the diver,—
One, when a beggar he prepares to plunge;
One, when a prince he rises with his pearl?
Festus, I plunge.

At lunch, I called her. 'The media has gone crazy,' she said. 'Across the globe, every channel is carrying it.' I found myself moved that our discovery meant so much to so many people. From out of the amber of another age, this wreck and its story seem to have touched everyone.

For me the day passed in a bit of a blur as one interview followed another. I struggled to remain fresh and interesting in the face of what were almost identical questions, but it seemed to go well. John and Dan were also hard at it, particularly Dan and his team, who for 12 hours never stopped. Back in London, Donald too was making TV and radio appearances and, of course, speaking to the press.

During the late afternoon I turned on my computer and was surprised by the number of goodwill messages I was receiving from all over the world. As might be expected, many were from friends and colleagues, but there were also a large number from people I did not know. Among the official letters of congratulations was one from the mayor of Cape Town to the Trust:

> Having sailed from Cape Town on a Cape Town-registered, South African vessel, the city is immensely proud of its association with your Expedition. Your discovery marks an important historical and scientific moment, not only for Cape Town but the whole world.

The messages that meant most to me came from the Falklands, starting with a message from the governor, who was delighted that the Falklands were so closely associated with this 'historic achievement'. A recurring theme was the relief that such a positive story brought from the harrowing reports coming in from the Russian invasion of Ukraine.

I passed Nico on the stairs and gave him a smile. What is there to say when you have just found the *Endurance*? And then he stopped. 'I have achieved my objective,' he said.

'That's good,' I replied.

Realizing that his allusion had escaped me, he explained: 'Do you remember what I said back in Gardanne?' I cast back. It seemed like an age ago. He went on, 'I said my objective on this project was to put a smile on your face.' We both laughed. He had sure done that.

The weather is deteriorating, but the ship is still on maximum revs and racing hell for leather towards South Georgia.

NOON POSITION: 64° 58.577' S, 045° 15.028' W

10 MARCH 2022

The death of Shackleton

There is storm weather pumping in from the west, but other than that, today was much like yesterday in that it was devoted to the media. Today it was the turn of the newspapers, which had all gone to press before yesterday's statement was released.

Since the epic inspection dive, I have been calling the giant white anemone over the 'AN' of the ship's name 'Annie' and the one presiding over the bow 'Ernie', after Shackleton, while the crab on the galley wall I was calling 'Wuzzles' after Worsley, the ship's captain. I said all this yesterday on one of the radio broadcasts; a print journalist must have been listening, because there it is today in one of the papers. Well, why not? Kids will love it, and it points toward the important matter of the biological communities living on the wreck.

When Shackleton got back to Britain after the *Endurance* expedition, he struggled. Conventional domesticity and the quiet life did not sit well with him and in 1920, if not before, he began talking about another polar expedition. His health was not good; he was smoking as much as ever, drinking heavily and putting on weight. As usual he was without money of his own and, perhaps not surprisingly, there were still outstanding debts from the *Endurance* expedition.

His first thought was the Arctic, but following some funding misfortunes with the Canadian government, his focus returned to Antarctica. With backing from an old school friend, John Quiller Rowett, who had made his fortune in rum, he set off from St

Katharine Docks beside Tower Bridge on 17 September 1921. The vessel he was using was a wooden sealer, *Foca I*, which he had bought in Norway and renamed *Quest*. She was an uncomfortable old tub with a cumbersome rig and dodgy engines. Shackleton's rather woolly objective was to circumnavigate the Antarctic Continent, searching for uncharted islands to explore, and perhaps conduct a survey of Enderby Land, a little-studied quadrant of the continental coastline that is today considered by Australia to be part of its Antarctic territory. His plans, if there were any, lacked definition.

The core of the team was what Shackleton called 'the old guard' from the *Endurance*, with Worsley once more as skipper. Frank Wild, of course, was there in his old role, as were his former doctors, Macklin and McIlroy. Hussey was back and still playing the same banjo, but having become a doctor, he was now also part of the medical staff. From the old crew came Alexander Kerr as engineer, Thomas McLeod was back on deck and Charlie Green was once more in the galley.

Almost straight away they all noticed that Shackleton was no longer the person he had been. His vitality had gone and Macklin realized he was not a well man. At Rio de Janeiro he suffered a heart attack but, typically, refused to let Macklin examine him. Their immediate destination was South Georgia but after that, wrote Macklin, 'he does not know what he will do'. This, indeed, was not the Shackleton of old.

Their voyage through the South Atlantic was so rough that they were obliged at times to heave to and pour oil on the waters to prevent being swept by breaking seas. (This ancient practice is the true origin of the expression 'pour oil on troubled waters'.) Even Shackleton got a bit seasick. Through night skies on 4 January 1922, 16 days out of Rio, they spotted the silhouetted mountains of South Georgia. When the sun rose, they could see the white-mantled mountains of the Allardyce Range. Worsley wrote of how he and Shackleton 'fought our battles o'er again' as they ran 'about the ship like two excited school-boys pointing out to each other and everyone on board the different spots we could recognize of our march across South Georgia'.

During the afternoon, in bright sunshine, they anchored in King Edward Cove at the whaling station of Grytviken. Frithjof Jacobsen, the old station manager, was still there, and Shackleton, who had shed his despondency, displayed a glint of his former self. Back on board that evening, he wrote in his diary: 'It is a strange and curious place. A wonderful evening. In the darkening twilight I saw a lone star hover gem-like above the bay.'

That night, at about a quarter to three, Shackleton was seized by acute chest pains.

In a letter to his wife, Jean, Worsley wrote:

Poor Shackleton died at 3 this morning of heart
disease . . . Last night he and I played cards, poker, patience,
up to 10 p.m. He was absolutely normal and cheerful and kept
on chafing me in his dear old way. Then we had a short yarn
about geological matters, future programmes, etc. After
setting the watches for the night I turned in. At about 2.35 he
whistled to Macklin who was then on watch and told him
he'd had a spasm of pain in his back and was then suffering
badly from another one, and requested Mac in his usual
imperious way to get him medicine to alleviate it. This Mac
did, but on returning found him very bad. He said 'You're
always telling me to give things up, Mac. What am I to give
up now?' in a childlike way. He then tried to take the
medicine but Mac fearing it was fatal called McIlroy but Sir E
died as the latter arrived about 2.15 a.m. . . . In the afternoon
we took Sir Ernest's body ashore and later the Norwegians
put it in a hermetically sealed tin coffin after our surgeons had
prepared the body against decay to the best of our resources.
We are sending the body to England in charge of Hussey.

Macklin had the unpleasant duty of performing a post-mortem on his old friend. The cause of death, he later wrote, was atheroma of the coronary arteries. Once Shackleton's body was sealed into the tin coffin described by Worsley, it was resolved to send his corpse back to England via Montevideo. South Georgia still did not have a radio, so, as no one knew what Lady Shackleton might wish, Hussey was deputed to accompany the body while Wild and Worsley decided to continue with the expedition. Once at Montevideo Hussey wired the expedition's supporter, John Quiller Rowett, and asked that he break the news to Shackleton's wife, Emily.

Emily decided that her husband should be buried at South Georgia. This would not have displeased Shackleton, as the island's great crags and deep fjords were more his spiritual home than anywhere else on earth. So, following a service at the Anglican church in Montevideo, his coffin was transported on a gun carriage to the quay, where it was put on board the vessel *Woodville* to go back, again with Hussey, to South Georgia.

They arrived on 27 February and, in a blinding snowstorm, the coffin was carried past the flensing decks to the wooden Lutheran church at the rear of the station. On 5 March, a service was held in English and Norwegian. The coffin was then carried on a light railway to the foot of Cemetery Hill, where the pall-bearers carried it up the rise to the open grave where the service was completed.

While waiting for a ship to carry him back to Montevideo, Hussey was surprised on 6 April to find the *Quest* back in harbour. Wild and the crew were equally startled by the appearance of Hussey. Before the *Quest* left South Georgia, they erected a stone cairn to Shackleton on the bluff at Hope Point which still stands to this day.

NOON POSITION: 59° 35.942' S, 039° 24.934' W

11 MARCH 2022

The *Endurance* left South Georgia on 5 December 1914.

For those on the *Endurance* as for us on the *Agulhas*, it was a moment for reflection. Not only did they dwell on what lay ahead, but they could not help wondering about how things were unfolding in what Orde-Lees called the 'vile war'. Although they did not know it, on that day the first shots were fired in what would become known as the Battle of the Marne.

The irony of it all, as we head towards South Georgia, is that we too cannot stop thinking about the current map-altering war in Europe, which for us also is 'the great national crisis of modern time'.

As we crept up Cumberland Bay to Grytviken, everybody was on deck. As always when coming into South Georgia, it was a tad mystical. The dark azure sea about us was waist-deep in mist, and before us, the great Allardyce massif was mantled in snow and wreathed in thick, low, linear cloud formations. As I watched, a great *whoosh* of loose vaporous snow was blown up and over the crest of Mount Sugartop from somewhere on the other side of the island, where the *James Caird* had once made landfall. All of us were agog, some a bit sloshy, and when we spoke it was from under our breath. It was a little like entering some great cathedral like Chartres or Notre-Dame de Paris.

I believe all of us, at that moment, were thinking about Shackleton. He means different things to different people. Everybody talks about his leadership, determination and courage, and yes, I get all that, but for me Shackleton has always been about how you live your life. Life, for him, was never just a journey to the grave. No; you squeeze it to the last drop, you suck the living daylights from it. Life is there to be grabbed by the scruff and then you run with it until, beaten and bruised, you drop. And that was Shackleton. He showed us how to live. In the words of his favourite poet, he said, 'Let me taste the whole of it.'

We landed by boat. I wandered about by myself. I had first been here when I worked at sea at the end of the 1960s. Grytviken was a very different place then. The station had only been closed for three or four years and was just as the whalers had left it, thinking they would be back the next season. But they had hooked too many leviathans; the bottom had dropped out of their industry, and they never came back. In those days the beaches were a bone-yard, the place stank of dead whale, and sheathbills and giant petrels crowded the shore waiting for the next feed that never came.

These days the beaches are clean, the scavenger gulls are gone and everything has been tarted up and neatened for the tourists. But for those of us who knew it in the '60s, it seems to have been destroyed by appalling ignorance and historic insensitivity.

I walked over to the Lutheran church where they held Shackleton's funeral service. Later, we all gathered to the east of the great flensing deck and then, quiet and sombre, we all processed along the track up to the cemetery on the knoll. Pipits the size of sparrows flitted back and forth, and the juvenile fur seals that were everywhere craned their necks for a better view of what we were doing.

Shackleton's grave is marked by a large granite pillar beside the white wooden fence at the back of the enclosure, opposite the entrance. All the graves are aligned east–west except for his, which faces south. On the back of his stone is a line from Browning: 'I hold that a man should strive to the utmost for his life's set prize.'

Our purpose was to hold a brief ceremony. It was to be filmed. Captain Knowledge stood at the east side of the headstone, I stood at the west, and then came John and Nico. I waited until everyone had gathered and then I sought the gaze of the film director, Natalie Hewit, who gave me a discreet nod to indicate that her cameramen were ready. I then called upon Captain Knowledge to speak first. Knowledge by nature is a quiet, private man, and we had not discussed his contribution, so I was curious to know what he would say and how he would say it. Without reference to notes, he extemporized one of the most heartfelt eulogies I have ever heard. He addressed Shackleton directly. 'Boss,' he said, 'I have come to tell you that we have found your baby.'

He was talking as one ship's captain to another. Then he spoke to us about Shackleton's effort to find and then raise the money to purchase the *Endurance*, and how she became his life and soul. He had with him a large photograph of the wreck on the seabed that we had all signed. Addressing Shackleton once more, he thanked him for guiding us and then said that he wanted to show him what his ship now looked like. He knelt down, and, like Shackleton laying Queen Alexandra's Bible on the snow at Dump Camp, he placed the photo on the grave.

I knew, of course, how much finding the *Endurance* meant to Knowledge, but until today I had not realized how much Shackleton himself meant to him. Later, when I was alone with Knowledge in his day room, he talked about how, when we were over the *Endurance*, he connected with Shackleton more than ever before. He thought particularly of Shackleton's leadership under stress and compared it to the challenges and dangers we had faced in 2019. He said, 'Finding the *Endurance* was the pinnacle of my career. I had to drive this ship and get us there but, of course, this project was so much bigger than me. And now we have succeeded, I feel that connection with Shackleton more than ever. Our discovery will revive the Shackleton name, story and legacy and make sure it never disappears into the history books.'

After Knowledge had delivered his eulogy I called on John, who spoke powerfully about the uncanny parallels between the war in Europe then and the one in Europe now. His words brought many of the team to tears.

And, because this was Shackleton, there had to be poetry. I had heard Dan Snow recite Browning's 'Prospice' before, so I invited him to read it again. It is a poem, or a monologue, about facing death, and how it should be without bandaged eyes nor fearing the night, for it is the moment when 'the worst turns the best'.

> Fear death? – To feel the fog in my throat,
> The mist in my face,
> When the snows begin and the blasts denote
> I am nearing the place . . .

It was one of Shackleton's favourites; he knew it by heart. He saw himself as its 'strong man' who 'must go'. Sometimes he would end his letters to his wife, Emily, with the word 'Prospice', which means 'to look forward'.

The survival of the *Endurance* as a great sealed box upon the seabed had caught everybody's imagination, and since the moment I had pointed out the portholes of Shackleton's cabin, everybody had been gripped by the thought of what might be inside. I knew Shackleton was able to save his case with all the expedition papers, but beyond that I was not sure of what might still be there. There was just one thing I knew for certain remained, and that was Rudyard Kipling's poem 'If'. Heavily framed and attached top and bottom with a single screw, it was midway between his two starboard portholes. So I read two verses from 'If' and then I spoke of the meaning of Shackleton. The film of our proceedings may show a wayward word or two, but I am writing this while it is fresh in my head, and this was what I said:

It's an old cliché that Shackleton never achieved anything he set out to do.

And it's true. He didn't.

But that is to miss the point. That is not what Shackleton was about.

He was about man's inherent urge to be always striding for the horizon, reaching for the next thing, pushing to expand his boundaries.

He exemplifies, better than anyone else I can think of, that thing within the human condition that sets us apart from the penguins, that thing that one day will carry us to the stars.

Shackleton may not have reached the Pole, but he blazed the trail for others to follow. By climbing the Beardmore Glacier and getting onto the Polar Plateau, he lifted the veil on the Great White Continent of Antarctica.

He could have reached the Pole. He could have claimed the prize. But he did not. He got to within a hundred miles of it. That is less than the length of this island. Why did he not do it? He didn't do it because he knew that if he did, on the way back men would die. And that is who Shackleton was.

As I stand here beside his grave with all of you, dear friends, crew and team, it occurs to me that in all Shackleton's expeditions into danger, which he himself led, the only life he lost was his own.

By 2330, the ship was secure for sea. On a heading of 33 degrees, we put out into 45-knot winds and five-metre waves. It was going to be a rough old night, but nobody cared, because coming here had been a perfect bookend.

NOON POSITION: 54° 42.948' S, 035° 31.789' W

EPILOGUE

Over the course of the following week, the ship worked her way back to Cape Town. I spent much of the time looking through the data sets and images the data processors had put together from the wreck. Some of the finds were astounding.

Pierre Le Gall told me that he thought he had found a handgun but needed to do a little more processing before he could be sure. I was a bit doubtful; I knew that the men of the *Endurance* had taken their pistols onto the ice with them. But then he showed me the results – and he was right. It was indeed a gun – a *flare* gun. This was a particularly exciting discovery, as it recalled a story told in the diaries about Shackleton, Wild and Hurley's last 'official' visit to the *Endurance* (other visits would be made right up until the day she sank). On that day, 8 November, they had fired a detonator using a gun. Hurley wrote: 'Pay the final official visit to the wreck with the Chief and Wild. Yesterday's mild blizzard has put the final touch to the destruction wrought by pressure and the salvage gangs . . . after saluting the ensign with a detonator fired on the poop, we returned sadly to the camp.'

In his diary, Shackleton wrote: 'Beautiful day . . . Hurley, Wild, Self, went on to ship, said Goodbye, fired a bomb in farewell.' Of course, we cannot be certain that this was the actual gun they fired, because vessels at that time always carried more than one.

And everybody wants to know about the ship's bell, and I think we have found it. I've mentioned before that I thought I could see the bell hanging from the beams in their winter quarters, or 'the Ritz', in one of the photographs. It must have been brought down with them to the interior of the ship from the bridge when they left their cabins in March 1915. But then, when they returned to their old cabins in October, the bell went back to its former position at the front of the bridge, where perhaps it was attached to the back of the foremast. With the destruction of the foremast, it apparently fell to its current position: within the gap between the mast and what is left of the forward wall of the deckhouse. More processing is required in order to be certain, but from what we can now see, there is something there that looks very much like the ship's bell.

And now I'm in the same bar at the same hotel in Cape Town where I sat three years ago. Once again, I'm with my old friend Chad Bonin. We are even in the same seats.

I have vivid memories of the last time we were here. We had taken such a drubbing. We sat feeling bemused by all that had happened, and not in a good way. Today, again, the feeling is one of bemusement, but this time it is all good. We have been lucky. The ice was kind, our technology worked and, in the end, we touched the Grail. It was an incredible team effort and everyone played a crucial role. But special mention must go to the subsea team, who delivered when it mattered.

It is way past midnight, but somehow neither of us wants to end the day. Although we have kind of said it all, we don't want it to be over. A few more jokes; a bit more tinpot wisdom. In a few hours I will be on the plane to England and he will be heading back to Louisiana.

'That wreck fairy sure left something under your pillow,' Chad laughs.

'Yeah,' I agree. 'A fairy-tale ending, you might say.'

'Whatever we do from now on, we are never going to be able to top this one.'

There is a long pause while we both consider this. I ponder the next chapters ahead of me, wondering whether anything could even compare.

'You know what gets me?' says Chad. 'I mean, about the whole Shackleton thing. It's that those guys got away with it. They shouldn't have. Twenty-eight went in and twenty-eight came out.'

I think about this for a bit. 'No,' I gently correct him, 'not twenty-eight. There were twenty-seven of them, and Shackleton.'

APPENDIX 1

Members of the Imperial Trans-Antarctic Expedition 1914–17

NAME (AGE AT START OF EXPEDITION)	POSITION	NATIONALITY
Sir Ernest H. Shackleton (40)	Leader	Ireland
Frank Wild (41)	Second-in-command	England
Frank Worsley (42)	Captain	New Zealand
Hubert T. Hudson (27)	Navigational officer	England
Lionel Greenstreet (25)	First officer	England
Thomas Crean (37)	Second officer	Ireland
Alfred Cheetham (48)	Third officer	England
Lewis Rickinson (31)	First engineer	England
Alexander J. Kerr (21)	Second engineer	England
Dr Alexander H. Macklin (25)	Surgeon	England
Dr James A. McIlroy (34)	Surgeon	England
James M. Wordie (25)	Geologist	Scotland
Robert S. Clark (31)	Biologist	Scotland
Leonard D. A. Hussey (23)	Meteorologist	England
Reginald 'Jimmy' W. James (23)	Physicist	England
James 'Frank' Francis Hurley (28)	Official photographer	Australia
George E. Marston (32)	Official artist	England
Thomas H. Orde-Lees (37)	Motor expert and storekeeper	England
Charles J. Green (25)	Cook	England
Harry 'Chippy' McNish (39)	Carpenter	Scotland
Walter Ernest How (28)	Able seaman	England
William Lincoln Bakewell (25)	Able seaman	United States

Timothy McCarthy (26)	Able seaman	Ireland
Thomas McLeod (40)	Able seaman	Scotland
John Vincent (30)	Able seaman	England
Ernest Holness (21)	Fireman	England
William Stephenson (25)	Fireman	England
Perce Blackborow (19)	Stowaway (later steward)	Wales

APPENDIX 2

Members of the Weddell Sea Expedition 2019

SENIOR EXPEDITION TEAM

John Shears	Expedition leader	United Kingdom
Mensun Bound	Director of exploration	Falkland Islands/United Kingdom
Professor Julian Dowdeswell	Chief scientist; director of the Scott Polar Research Institute (SPRI), University of Cambridge	United Kingdom
Channing Thomas	AUV and subsea team leader, Ocean Infinity (OI)	United States
Claire Samuel	Offshore manager, Deep Ocean Search Ltd (DOS)	France
Captain Knowledge Bengu	Master of S.A. *Agulhas II*	South Africa
Captain Freddie Ligthelm	Ice pilot	South Africa
Dr Claire Grogan	Expedition doctor	United Kingdom

SCIENTISTS AND ENGINEERS

Professor Sarah Fawcett	Oceanographer and chief scientist, University of Cape Town (UCT)	South Africa
Katherine Hutchinson	Oceanographer, UCT	South Africa
Tahlia Henry	CTD operator and oceanographer, UCT	South Africa
Riesna Audh	Oceanography PhD student, UCT	South Africa
Harry Luyt	Oceanography PhD student, UCT	South Africa

Jessica Burger	Oceanography PhD student, UCT	South Africa
Raquel Flynn	Oceanography PhD student, UCT	South Africa
Shantelle Smith	Oceanography MSc student, UCT	South Africa
Kurt Spence	Oceanography MSc student, UCT	South Africa
Evelyn Dowdeswell	Geologist, SPRI	United Kingdom
Christine Batchelor	Geophysicist, SPRI	United Kingdom
Frazer Christie	Glaciologist and remote sensing analyst, SPRI	United Kingdom
Aleksandr 'Sasha' Montelli	Glacial geology researcher, SPRI	Russia
Professor Anriëtte Bekker	Mechanical engineer, Sound and Vibration Research Group, Stellenbosch University	South Africa
James-John Matthee	Mechanical engineering MEng student, Stellenbosch University	South Africa
Christof van Zijl	Mechanical engineering MEng student, Stellenbosch University	South Africa
Tommy Bornman	Elwandle Coastal Node manager, South African Environmental Observation Network	South Africa
Jeff Evans	Senior lecturer in physical geography, University of Loughborough	United Kingdom
Liangliang Lu	Naval architect, Marine and Arctic Technology Research Group, Aalto University, Finland	China
Dag Ottesen	Geologist, Norwegian Geological Survey	Norway
Wolfgang Rack	Glaciologist, University of Canterbury, New Zealand (UC)	Germany

Paul Bealing	Drone operator, UC	New Zealand
Michelle Taylor	Marine biologist, University of Essex	United Kingdom
Lucy Woodall	Senior marine biologist, University of Oxford	United Kingdom
Betina Frinault	Marine biology DPhil student, University of Oxford	France/United Kingdom
Leon Wuis	Senior sea technician, Royal Netherlands Institute for Sea Research	Netherlands

AUV AND SUBSEA TEAMS

Chad Bonin	AUV technician, OI	United States
Blake Howard	AUV technician, OI	United States
Devon James	AUV technician, OI	United States
Todd Oxner	AUV technician, OI	United States
Espen Strange	Senior AUV technician, Kongsberg Maritime	Norway
Pierre Le Gall	Survey data analyst, DOS	France
Julien Trincali	Survey data analyst, DOS	France
Steven Saint Amour	ROV team leader, Eclipse	United States
Ray Darville	ROV technician, Eclipse	United States
Steven March	ROV pilot, Eclipse	United States
Dave O'Hara	ROV pilot, Eclipse	United Kingdom

OTHER SPECIALISTS

Holly Ewart	Project manager, OI; representative of the Flotilla Foundation	United Kingdom
Colin de la Harpe	Team leader, South African Council for Industrial and Scientific Research	South Africa
Tim Huseyin	IT technician, OI	United Kingdom

| Thapi Magabutlane | Meteorologist, South African Weather Service | South Africa |

DOCUMENTARY TEAM

Olive King	Location director	United Kingdom
Paul Williams	Cameraman	United Kingdom
Kobus Loubser	Assistant cameraman	United Kingdom
Tamara Stubbs	Sound technician and drone operator	United Kingdom

APPENDIX 3

Members of the Endurance22 Expedition 2022

SENIOR EXPEDITION TEAM

John Shears	Expedition leader	United Kingdom
Mensun Bound	Director of exploration	Falkland Islands/United Kingdom
Nico Vincent	Expedition subsea manager, Deep Ocean Search Ltd (DOS)	France
Dr Lasse Rabenstein	Chief scientist; Drift+Noise Polar Services	Germany
Jean-Christophe 'JC' Caillens	Offshore manager, DOS	France
Carl Elkington	Ice camp manager, White Desert	South Africa
Michiel Swanepoel	Helicopter operations manager, Ultimate HELI	South Africa
Natalie Hewit	Documentary director, Little Dot Studios	United Kingdom
Captain Knowledge Bengu	Master of S.A. *Agulhas II*	South Africa
Captain Freddie Ligthelm	Ice pilot	South Africa
Dr Lucy Coulter	Expedition doctor	United Kingdom

SCIENTISTS AND ENGINEERS

Marc De Vos	Senior meteorologist / oceanographer, South African Weather Service	South Africa
Carla-Louise Ramjukadh	Meteorologist / oceanographer, South African Weather Service	South Africa

Professor Jukka Tuhkuri	Senior ship engineering scientist; professor in solid mechanics at Aalto University	Finland
Christian Katlein	Sea ice scientist, Alfred Wegener Institute for Polar and Marine Research	Germany
Stefanie Arndt	Sea ice scientist, Alfred Wegener Institute	Germany
Jakob Belter	Senior sea ice scientist, Alfred Wegener Institute	Germany
Beat Rinderknecht	Science technician, Drift+Noise	Switzerland
Alexandra Stocker	Sea ice science student, Drift+Noise	Switzerland
Mira Suhrhoff	Sea ice science student, Drift+Noise	Germany
Thomas Busche	Senior remote sensing manager, German Aerospace Center	Germany
Dmitrii Murashkin	Remote sensing sea ice specialist, German Aerospace Center	Russia
Anriëtte Bekker	Mathematical statistician, Stellenbosch University	South Africa
Ben Steyn	Ship engineering MSc student, Stellenbosch University	South Africa
James-John Matthee	Mechanical engineering scientist, Stellenbosch University	South Africa
John Albertson	Marine archaeologist, SEARCH Inc.	United States

AUV AND SUBSEA TEAMS

Chad Bonin	AUV supervisor, Ocean Infinity (OI)	United States
Robbie McGunnigle	AUV pilot and technician, OI	United Kingdom
Joe Leek	AUV pilot and technician, OI	United Kingdom
Maeva Onde	Senior surveyor, DOS	France

Clément Schapman	Senior surveyor, DOS	France
Pierre Le Gall	Data processor, DOS	France
François Macé	Senior surveyor, DOS	France
Fred Soul	Data processor, DOS	France
Jérémie 'Jim' Morizet	Subsea survey engineer, DOS	France
Kerry Taylor	AUV supervisor, DOS	United Kingdom
Grégoire Morizet	Senior surveyor, DOS	France
Thomas Andreasson	AUV engineer, SAAB	Sweden
Lars Lundberg	AUV engineer, SAAB	Sweden

AVIATION TEAM

Warren Vogt	Helicopter engineer	South Africa
Waldo Venter	Helicopter pilot	South Africa
Zakaria Johnson	Helicopter ground team	South Africa
Darius Carstens	Helicopter avionics engineer	South Africa
Eduan Teich	Helicopter engineer	South Africa
Jodi Brophy	Helicopter engineer	South Africa
Charles Tait	Chief helicopter pilot	South Africa
Scott Barnes	Helicopter pilot	South Africa
Buzz Bezuidenhout	Helicopter pilot	South Africa
Timothy Hughes	Helicopter engineer	United States
Michael Patz	Helicopter engineer	United States
Chad Halstead	Helicopter pilot	United States
Kevin Brashar	Helicopter pilot	United States

WHITE DESERT ICE CAMPS TEAM

Emmanuel Guy	Ice camp first aider	France
Grant Clark	Field guide	South Africa
Tom Ross	Field guide	South Africa
Chad Burtt	Field guide	South Africa

Grant Brokensha	Field guide	South Africa
Wayne Auton	Field guide/paramedic	United Kingdom
Nicholas Burden	Camp leader	South Africa

MEDIA AND EDUCATIONAL OUTREACH

Dan Snow	History broadcaster	United Kingdom
Esther Horvath	Expedition photographer, Alfred Wegner Institute	Hungary
Nick Birtwistle	Producer, Little Dot Studios	United Kingdom
James Blake	Aerial cameraman, Little Dot	United Kingdom
Paul Morris	Producer, Little Dot	United Kingdom
Saunders Carmichael-Brown	Presenter, Little Dot	United Kingdom
Tim Jacob	Education outreach coordinator, Reach The World	United States
Frédéric Bassemayousse	Expedition photographer	France

ACKNOWLEDGEMENTS

Our first search for the *Endurance* took place in 2019 and was called the Weddell Sea Expedition (hereafter WSE); the second search occurred in 2022 and was called Endurance22. Whilst there were differences between the two seasons in management, team composition, subsea technology and general methodology, there were also many consistencies between them and certainly the success of Endurance22 was built upon hard lessons learned in 2019. In every sense these campaigns were a team effort, and the gratitude that is owing to them all individually, and as an assembly of experts *in toto,* is beyond words. In what follows I shall record first the names of those on shore whose efforts made the work at sea possible, after which I will focus on the main figures in authority off-shore and within the field teams before turning to the various organizations that gave academic or practical assistance, and then, finally, I will name those individuals who helped in other ways, but who do not necessarily fit within the foregoing groupings.

All of us involved in both WSE and Endurance22 owe a great debt to the private and understated man whose tremendous passion for ocean exploration was the inspiration for the project. We are next most indebted to the team at Ocean Infinity (hereafter OI), the deep-water robotics company that, during both campaigns, provided the submersibles and their technical support teams. Without OI the search would not have happened – it's as simple as that. The management committee for WSE was made up of Gwilym Ashworth, George Horsington, John Kingsford, Donald Lamont, John Shears and myself. The management of the 2022 campaign was overseen by the trustees of the Falklands Maritime Heritage Trust (FMHT), and here special mention must be made of its chairman, Donald Lamont, a distinguished former governor of the Falklands who bestrode every aspect of Endurance22; for three years he worked without respite or reward through a period of exceptional difficulty when, because of the Covid-19 pandemic, the world was in lockdown. The debt of all on the project to Donald cannot be overstated. The other trustees of the FMHT are Bill Featherstone, Saul Pitaluga and myself. Bill and Saul are deserving of special praise as their dedication and hard work was the bedrock upon which Endurance22 was built. During planning and general organization for 2019 and 2022, two other members of the shore-based staff stood out for the scale and excellence of their contribution: they were George Horsington (ABC Maritime AG) and Sébastien Bougant of Deep Ocean Search (hereafter DOS). Another to whom we owe thanks is Bob Ormerod (DOS) who oversaw mobilization for WSE. In the long run-up to 2019, Prof. Julian Dowdeswell, then director of the Scott Polar Research Institute at Cambridge, worked hard to assemble an outstanding team of polar scientists; I only wish there had been space in this book to cover their important work in the detail it deserves.

Turning now to the two field seasons and their teams. Here my foremost debt is to John Shears (Shears Polar) who was the expedition leader of both campaigns and was in every way exceptional in his duties as well as in his attention to everyday detail and the care of the team and its safety. Special acknowledgement must also be recorded of the outstanding contributions made in 2019 by Channing Thomas (OI) who was in charge of the AUVs and all subsea activity, Claire Samuel (DOS) who ran the back deck, Holly Ewart who was the OI project manager, and Steve Saint Amour who supervised all ROV operations; to have been able to work alongside such a group of complete professionals was a privilege. In 2022 particular gratitude is owing to JC Caillens (OI/DOS) who oversaw all back-deck activity and Nico Vincent (DOS) who was manager-in-charge of all equipment preparation, testing and training ashore and all subsea operations and related topside activities at sea. On a project like this, it is always delicate to single out one person for special praise, but I do not think that there is anybody on Endurance22 who would not agree that, in achieving our goal, we owe a huge debt of gratitude to Nico who, so to speak, designed the rocket ship and then landed it.

Finally, with regard to the field seasons, I must also highlight the important contributions made by the medical officers (Claire Grogan, 2019; and Lucy Coulter, 2022) and the chief scientists (Prof. Julian Dowdeswell, 3 January to 2 February 2019; Prof. Sarah Fawcett, 4 February to 20 February 2019; Dr Lasse Rabenstein, 2022). The photographers whose work graces these pages and our website were Colin de la Harpe, Esther Horvath, Fréd Bassemayousse, Pierre Le Gall and Julien Trincali. During both campaigns all public relations matters were overseen by Celicourt Communications. Here our foremost debt of gratitude is to Mark Antelme, ably assisted by his team. In all things Mark worked closely with Alex Wright of AW Design, who with great patience and expertise ran the project website. Our endeavours in the Weddell Sea were conducted under a permit from the Polar Regions Department of the Foreign and Commonwealth Office and here we are grateful for the support of Jane Rumble, George Clarkson, Ariane Watson and Thomas Chance.

Major successes of both 2019 and 2022 were the educational outreach programmes that were conducted through close ties with Reach the World (hereafter RTW), and media partners History Hit, Little Dot Studios and Consequential. Our involvement with RTW, which brought us into direct contact with schoolchildren and their teachers, involved many people, but my leading expression of gratitude must be to Julianne Chase, who was the first to spot the potential benefits of a collaboration between WSE and RTW. She enrolled the support of John Kingsford, who made the necessary representations to the management committee who immediately saw the merits and embraced the partnership. Our association with RTW took us into classrooms around the world and enabled us to talk directly to the young about what we were doing and thus – we hope – inspire them to pursue an interest in science, the stewardship of the environment and exploration-related careers in general.

During 2022, in over 32 livestream events (the majority of which were from the ship in the Weddell Sea), we were joined by more than 33,000 students in 28 countries around the world. The educational programme designed by RTW lasted a year and involved 61 curriculum-related papers to assist educators with classroom learning and special projects. Within RTW it is my pleasure to be able to record our debt to Heather Halstead, Timothy Jacob, Colin Teague, Christopher Ahearn, Ziev Dalsheim-Kahane, Brianna Rowe and Holly Ewart of OI who was the 2019 on-board presenter. Other organizations that supported or collaborated with RTW were Exploring by the Seat of Your Pants (Joe Grabowski), the Explorers Club and the Royal Geographical Society.

In addition to students, FMHT was also keen to reach a wider global public and this was achieved through History Hit, which released regular podcasts and made daily transmissions across a range of media platforms. Within History Hit our foremost thanks goes to its founder, the historian Dan Snow. Whatever the temperature, weather or time of day, Dan was always there reporting to the world on the search and, of course, when it at last came, the moment of discovery. Another important aim of Endurance22 was the filming of a documentary and this was achieved by a team led by Little Dot Studios, History Hit and Consequential with the indefatigable and extremely talented Natalie Hewit at the helm. While on the subject of filming, I would like also to record our gratitude to Anthony Geffen, whose company Atlantic Productions, under Olive King and Oliver Twinch, made an excellent documentary on the 2019 campaign (Endurance: *The Hunt for Shackleton's Ice Ship*) which was broadcast by the History Channel in its History's Greatest Mysteries series.

I now come to the various museums, archives, libraries and learned societies that contributed to my research, starting in the Falkland Islands where it is my particular pleasure to be able record my gratitude to the chairman of the trustees at the Falkland Islands Museum, Richard Cockwell, and the museum director Andrea Barlow, as well as her associates Teena Ormond, Sandra Alazia,

Shirley Hirtle, Tasmin Tyrrell and Tara Hewitt. Tansy Bishop of the Jane Cameron National Archives helped with research. Important support also came from the former governor of the Falklands, Air Commodore Nigel Phillips. In England my greatest debt is to the archive, library and museum of the Scott Polar Research Institute, Cambridge, where I spent many days going through diaries and other documents seeking information that might help me draw up our search box. At SPRI the two who assisted me most were Naomi Boneham and Laura Ibbett in the archive department. Others who gave much-appreciated aid were Prof. Julian Dowdeswell, Prof. Neil Arnold, Charlotte Connelly, Lucy Martin, Peter Lund and Nicola Hudson. I also received much help from the Shackleton Museum at Athy (Seamus Taaffe, Kevin Kenny, Sinead Cullen). Other welcome support came from archivists, librarians and curators at the National Archives, Kew; the National Maritime Museum, Greenwich (Simon Stephens, Jeremy Mitchell, Andrew Choong); the Alexander Turnbull Library, Wellington, New Zealand; the National Library of Australia; Dartmouth College Library, New Hampshire; the Norwegian Maritime Museum, Oslo (Elisabeth S. Koren); the Whaling Museum, Sandefjord (Dag Ingemar Børresen); the Vestfold Museums, Sandefjord; and the Fram Museum, Oslo (Geir O. Kløver). In Norway I was particularly grateful to Stig-Tore Lund, without whose advice and guidance the quality of my research would have been much the poorer.

I received much encouragement and research assistance from some of the descendants of those on the Imperial Trans-Antarctic Expedition, starting with the granddaughter of Sir Ernest Shackleton, the Hon. Alexandra Shackleton. The grandchildren of James Wordie, Pippa and Roderick Wordie, very kindly lent me a copy of their grandfather's diary and other materials which I took with me on Endurance22. Viv and John James, the sons of Reginald James, kindly made available to me their father's diary and copies of his lectures. Finally, with regard to research, there has to be a special mention of Stephen Scott-Fawcett, the pre-eminent Shackleton scholar of our times, and the one to whom I always turned for help with the more recondite corners of the Shackleton story. I am also indebted to Stephen for allowing me to report on the progress of the search to members of the Sir Ernest H. Shackleton Appreciation Society, the largest and most respected community of its kind in the world.

Ever since the launch of the search for the *Endurance* at the Royal Geographic Society (RGS) in 2018 we have worked in close association with the Society, particularly in 2019 when the Flotilla Foundation collaborated with the RGS to hold a conference on the scientific findings of the 2019 campaign, which was organized by Melanie Smith, ably assisted by Holly Ewart. At the RGS we are especially indebted to Prof. Joe Smith (Director) and his associates Steve Brace, Christine James, Alasdair MacLeod, Laura Melville, Lucy Preston, Nigel Winser and Shane Winser. Mention might also here be made of the All-Party Parliamentary Group for the Polar Regions (James Gray MP, Sophie Montagne) that twice hosted presentations of our work at Portcullis House.

Since 2018 the project has worked in close association with the Explorers Club of New York (of which Shackleton was a member) which organized talks and presentations to celebrate the search and its eventual success. During both campaigns the author was honoured to be made flag-bearer for the Club. Particular gratitude is owing to the current president of the Club, Richard Garriott de Cayeux, as well as to ex-president Richard Wiese, Julianne Chase, David Isserman, Martin Kraus, Mike Massimino, Kevin Murphy, Martin Nweeia, Ann Passer, Will Roseman and Angela Schuster.

Turning now to the operational side of the project. It might be said that the two standout heroes of my story were ships: the *Endurance* and the S.A. *Agulhas II*. Of both I cannot speak more highly, but of the latter let me say that, in my view, and writing as one who has spent much of his life on vessels of one kind or another, there is no better working ship afloat and none, whether ice ship or otherwise, that is better managed and crewed. In this regard special mention must be made of the owners of the *Agulhas II*, the South African Department of Fisheries, Forestry and the Environment, and AMSOL (African Marine Solutions) that, with remarkable thoroughness and efficiency, operates the ship for the government. At AMSOL I particularly wish to mention project manager Dave Murray, vessel superintendent Rob Hales and CEO Paul Maclons. As I hope might be inferred from my narrative, while on the *Agulhas II* we enjoyed a particularly productive working relationship and, indeed, friendship with the vessel's master, Captain Knowledge Bengu, and ice pilot, Captain Freddie Ligthelm. Special thanks must also go to Jacques Walters, the chief officer in 2019, a giant of a character and one of the most remarkable seamen I have ever met. Others, who it was my privilege to have known, include the bosun Lionel Alexander, the IT officer Orlando February, and captains Regan Paul and Michael Mdluli. I would like to record here my special thanks to Deep Ocean Search (DOS), in particular its founder and CEO John Kingsford who, in 2017, was one of the first two people to be involved in project planning, and later became one of the small committee that oversaw all aspects of WSE. In addition, both campaigns were staffed by many of his company's hydrographers, surveyors, data analysts and subsea technicians.

A mainstay of Endurance22 was Drift+Noise, which helped predict the direction and speed of the ice; this was of fundamental importance in determining where to dive next and how best to position the ship. Our foremost debt is to Lasse Rabenstein, ably assisted in Germany by Panagiotis Kountouris as well as Bernhard Schmitz and Paul Cochrane of the IcySea information service. From the Alfred Wegener Institute we are grateful to Helge Goessling and Valentin Ludwig for SIDFEx ice drift forecasts, Stefan Hendricks for CryoSat-2 ice thickness data and Thomas Krumpen for logistical and administrative support. For satellite imagery we are grateful to Thomas Busche, Dmitrii Murashkin, Anja Frost and Egbert Schwarz from the German Aerospace Center (DLR); Michael Wollersheim from Iceye; Thomas König and Christine König from Dr Thomas König & Partner, Fernerkundung GbR; Jack Hild from Hild Enterprises; Megan Zaroda from Planet; and Gene C. Feldman and Norman Kuring from NASA.

The project was also much indebted to Ultimate HELI (Pty) Ltd of Midrand, Johannesburg (CEO Shaun Roseveare) for providing the Bell 412 utility helicopter and team, and to ROTAK Helicopter Services of Anchorage, Alaska, who at the last minute were able to provide a Kaman K-MAX intermeshing-rotor helicopter and team to undertake the project's heavy-lift requirements. Particular gratitude is owing to White Desert (Patrick Woodhead, Luke Brauteseth) who in 2019 flew us into Wolf's Fang, Queen Maud Land, and then on to Penguin Bukta on the Fimbul ice shelf to join the S.A. *Agulhas II*. In 2022 they dispatched a party, under Carl Elkington, to join the team in order to equip and run ice camps but, when this service was not required, they provided care for the back-deck team when temperatures dropped to a highly dangerous −40°C. Other firms and businesses that assisted include Ambris LLP insurance, Constantia Consulting, Eclipse Group ROV (Steve Saint Amour), Pitmans LLP legal advisors, the Shackleton Design and Manufacturing Company (Ian Holdcroft and Martin Brooks), Shackleton Whisky of Whyte and Mackay (Bryan Donaghey, Kieran Healey-Ryder, Carla Santoni) and Voyis Imaging Inc.

At a more personal level there are many people and friends who were not necessarily part of any organization but who helped in more discreet ways. To them all I extend heartfelt thanks. In alphabetic order they were: Dalya Alberge, Jonathan Amos, Chris Andrews, Fréd Bassemayousse, Toby Benham, Chad and Heidi Bonin, Mike and Sigi Critchley, Colin de la Harpe, Tim Dingemans, Jennifer Erwitt, Elizabeth Featherstone, Henry Fountain, Mark Frary, Neil Gilbert, Simon Gilbert, Richard and Claudia Green, Lindsey Hardwick, Richard Holme, Beverley Humphrey, Adam Iscoe, Zaahir Kaffoor, Sean Kingsley, Julie Lacroix, Lynda Lamont, Pierre Le Gall, Gus Meikle, Johny Midnight, Ollie Mills, Peter Morse, Carlos and Mariela Muñoz Nuñez, Dave O'Hara, Hugo Pickering, Caroline Pitaluga, Oliver Plunkett, Elizabeth Rabett, Debbie Robinson, Julian Sancton, Liz Shears, Libby Smith, Max Smith, Melanie Smith, Will Smith, Rick Smolan, Shawaal Sonday,

Kurt Spence, Peter Stanford, Tamara Stubbs, Rudiger Sturm, Charles Thompson, Julien Trincali, Laurence Turner, Ollie Twinch, Mark Wood, Max Worrin, Matthew Wortman and Taylor Zajonc.

And finally, in the production of this book, my first thought is to Luigi Bonomi (LBA Books), a determined and exceptionally astute agent who, since we first met, has become a friend. At Pan Macmillan I was lucky to work with two of the most brilliant editors I have ever known, Mike Harpley and Mike Jones, who questioned every comma and together crafted my rather inflated manuscript into something leaner and more readable. Others at Macmillan to whom I would like to express my thanks include James Annal, Sian Chilvers, Hannah Corbett, Samantha Fletcher, Nicole Foster, Jiri Greco, Rebecca Lloyd, Lindsay Nash, Molly Robinson, Camilla Rockwood, Holly Sheldrake and Natasha Tulett. I am also most grateful to Donald Lamont, Nico Vincent and John Shears who read the early text and caught many errors. My sons Cody, Mensun (Jr) and Zak did much to help and encourage me with the preparation of the first draft. But above all I could not have done this without my wife, Jo, who has worked with me on wreck excavations all around the world since we were students in the late 1970s. This book was her idea and at every page I sought to follow her seminal advice, which was to 'keep it in the moment and bring it back to Shackleton'.

TEXT CREDITS

Diaries of Frank Hurley reproduced from *The Diaries of Frank Hurley: 1912–1941* (Anthem Press, 2011).

Diaries of Leonard Hussey reproduced from Hussey, Leonard, *South with Shackleton* (Sampson Low, 1949).

Diaries of Reginald James reproduced with kind permission of the James family.

Diaries of Alexander Macklin: excerpts appear by permission of the University of Cambridge, Scott Polar Research Institute.

Diaries of Harry McNish © National Library of Australia, Canberra.

Diaries of Thomas Orde-Lees reproduced from Thomson, John (ed.), *Lost: The Antarctic Diaries of Thomas Orde-Lees* (Erskine Press, 2020).

Diaries of Ernest Shackleton reproduced from Shackleton, Ernest, *South* (William Heinemann, 1919).

Diaries of James Wordie reproduced with kind permission of the Wordie family.

Diaries of Frank Worsley: excerpts appear by permission of the University of Cambridge, Scott Polar Research Institute.

PICTURE CREDITS

First plate section

1. Royal Geographical Society/Alamy Stock Photo
2. Julien Trincali
3. Esther Horvath & Falklands Maritime Heritage Trust
4. Colin de la Harpe
5. Frank Hurley/Royal Geographical Society via Getty Images
6. Ibid.
7. Ibid.
8. Frank Hurley/Scott Polar Research Institute, University of Cambridge/Getty Images
9. Ibid.
10. Colin de la Harpe
11. Ibid.
12. Ibid.
13. Mensun Bound
14. Pictorial Press Ltd/Alamy Stock Photo
15. Science History Images/Alamy Stock Photo
16. Frank Hurley/Royal Geographical Society via Getty Images
17. History and Art Collection/Alamy Stock Photo
18. James-John Matthee
19. Frank Hurley/Scott Polar Research Institute, University of Cambridge/Getty Images
20. Colin de la Harpe

Second plate section

1. James Blake & Falklands Maritime Heritage Trust
2. Nick Birtwistle & Falklands Maritime Heritage Trust

3. Esther Horvath & Falklands Maritime Heritage Trust
4. Ibid.
5. Ernest Shackleton/Royal Geographical Society via Getty Images
6. Esther Horvath & Falklands Maritime Heritage Trust
7. Frank Hurley/Royal Geographical Society via Getty Images
8. Frank Hurley/Scott Polar Research Institute, University of Cambridge/Getty Images
9. Esther Horvath & Falklands Maritime Heritage Trust
10. Ibid.
11. Hulton Archive / Getty Images
12. Falklands Maritime Heritage Trust
13. Ibid.
14. Ibid.
15. Mensun Bound
16. Colin de la Harpe

SELECTED SHACKLETON BIBLIOGRAPHY

Alexander, Caroline, *The* Endurance: *Shackleton's Legendary Antarctic Expedition* (Alfred A. Knopf, 1998)

Bakewell, William Lincoln, *The American on the* Endurance: *Ice, Seas, and Terra Firma* (Dukes Hall Publishing, 2004)

Baughman, T. H. and Rosove, M. H. (eds.), *Rejoice My Heart: The Making of H. R. Mill's* The Life of Sir Ernest Shackleton (Adélie Books, 2007)

Begbie, Harold, *Shackleton: A Memory* (Mills & Boon, 1922)

Bickel, Lennard, *In Search of Frank Hurley* (Macmillan Australia, 1980)

Bickel, Lennard, *Shackleton's Forgotten Men: The Untold Tragedy of the* Endurance *Epic* (Pimlico, 2001)

Butler, Angie, *The Quest for Frank Wild* (Jackleberry, 2011)

Butler, Angie and Riffenburgh, Beau, *Shackleton's Critic: The Life and Diaries of Eric Stewart Marshall* (Jackleberry, 2020)

Cockram, Roy, *The Antarctic Chef: The Story of the Life of Charles Green* (Southampton, 1999)

Credland, Arthur G. (ed.), *Charles Green: The Antarctic Chef* (East Yorkshire Local History Society, No. 60, 2014)

Dunnett, Harding McG., *Shackleton's Boat; The Story of the* James Caird (Neville & Harding, 1996)

Fisher, Margery and Fisher, James, *Shackleton* (Barrie Books, 1957)

Huntford, Roland, *Shackleton* (Abacus, 1996)

Hurley, Frank, *Shackleton's Argonauts* (Angus & Robertson, 1948)

Hussey, Leonard, *South with Shackleton* (Sampson Low, 1951)

Jara, Mauricio F., Mancilla, Pablo G., *Rescate en la Antártica: Comisió Piloto 2° Luis Pardo Villalón en 1916; Héroe Popular* (LW Editorial, Fundación Valle Hermoso, 2019)

Lansing, Alfred, *Endurance: Shackleton's Incredible Voyage* (Hodder & Stoughton, 1959)

Larson, Edward J., *An Empire of Ice: Scott, Shackleton, and the Heroic Age of Antarctic Science* (Yale University Press, 2011)

McElrea, Richard and Harrowfield, David, *Polar Castaways: The Ross Sea Party (1914–17) of Sir Ernest Shackleton* (McGill-Queen's University Press, 2004)

Mill, Hugh Robert, *The Life of Sir Ernest Shackleton* (William Heinemann, 1923)

Mills, Leif, *Frank Wild* (Caedmon of Whitby, 1999)

Piggott, Jan (ed.), *Shackleton: The Antarctic and* Endurance (Dulwich College, 2000)

Richards, R. W., *The Ross Sea Shore Party, 1914–17* (Scott Polar Research Institute, 1962)

Shackleton, Ernest, *The Heart of the Antarctic* (William Heinemann, 1910)

Shackleton, Ernest, *South* (William Heinemann, 1919)

Smith, Michael, *An Unsung Hero: Tom Crean – Antarctic Survivor* (Collins, 2009)

Smith, Michael, *Polar Crusader: A Life of Sir James Wordie* (Birlinn, 2007)

Smith, Michael, *Shackleton: By Endurance we Conquer* (Oneworld Publications, 2014)

Summers, Julie, *The Shackleton Voyages* (Weidenfeld & Nicolson, 2002)

Taaffe, Seamus, *Nimrod: The Journal of the Ernest Shackleton Autumn School* (Athy Heritage Centre–Museum, 2008)

Thomson, John, *Elephant Island and Beyond: The Life and Diaries of Thomas Orde-Lees* (Bluntisham Books, 2003)

Thomson, John, *Shackleton's Captain: A Biography of Frank Worsley* (Hazard Press, 1998)

Tyler-Lewis, Kelly, *The Lost Men: The Harrowing Story of Shackleton's Ross Sea Party* (Bloomsbury, 2006)

Webster Smith, Bertie, *Scott of the Antarctic* (Blackie & Son, 1955)

Wild, Frank, *Shackleton's Last Voyage* (Cassell & Co., 1923)

Wöppke, C. L. and Jara, M. F., *El Piloto Luis Pardo Villalón: Visiones desde la prensa, 1916* (LW Editorial, Fundación Valle Hermoso, 2016)

Worsley, Frank, *Endurance: An Epic of Polar Adventure* (P. Allan, 1931)

Worsley, Frank, *Shackleton's Boat Journey* (P. Allan, 1933)

Wright, Joanna (et al.), South with Endurance: Shackleton's Antarctic Expedition 1914–1917 (Book Creation Services, 2001)

INDEX

acidification, ocean 27–8, 29

Admiralty Chart 4024 87, 112, 211, 216

Agulhas II, S.A.; see also Endurance22
Expedition (2022); Weddell Sea
Expedition (2019)
ability to withstand storms 194–5
'Annie,' 'Ernie' and 'Wuzzles' 364
cabins 138–9, 209
crew 10, 12–13
design, Ligthelm's involvement in
145
FRC (fast rescue craft) 34, 36, 50
helideck **Plate I.3**, 210–11
icebound **Plate I.4**, 139–41, 158, 166,
167–8, 169, 170, 174, 175, 270–1
icebreaking **Plate II.1**, 30, 51–2,
53–4, 91, 117, 144, 164, 168, 170–1,
172, 183–4, 261–2, 287
monkey island 17, 200
moon pool **Plate I.10**, 34, 157, 159,
161, 171, 251–2, 262
normal services 10, 118
photographs **Plate I.2**, **Plate I.3**,
Plate I.4, **Plate I.10**, **Plate I.12**,
Plate I.13, **Plate I.18**, **Plate II.1**

albatrosses 83, 242–3

Albertson, John 384

Andreasson, Thomas 385

anemometers 335–6

Antarctic 33, 86

Antarctica
the Big Spread 175–6
climate change impacts on 27–8,
29, 45, 71, 98

flash freezes 175
ice *see* ice, types of
no straight lines through ice 87–8,
111, 210, 261
Shackleton's sketch of **Plate II.5**
silence 139–40
tourism 97–8, 102
whaling industry 26–7
wildlife *see specific group*

Arndt, Stefanie (Steffi) 232–3, 384

Audh, Riesna 379

Aurora 3

Austin, Captain Horatio 146

Auton, Wayne 386

AUVs (autonomous underwater
vehicles)
altitude 150
AUV 7
giving up search for 177
going AWOL 160–6, 171–2, 174,
180
hopes for 182, 202
on missions 74, 83, 148–52, 154,
155, 158–60, 325
sledge flags 146
tests 63, 107, 117, 147–8
AUV 9 **Plate I.13**, 40, 42, 47–57,
70–1, 83, 151–2
HiPAP communication with 34,
53, 157, 159, 162, 163, 171, 174, 180
HiSAS 151–2
PMA (post-mission analysis) 150–1
side-scan sonar 150–1, 213
stingers 148, 197

Bakewell, William Lincoln (Bakie) 215, 377
bannocks 185–6
Barnes, Scott 385
Bassemayousse, Frédéric 266, 268, 302, 326, 348–9, 386
Batchelor, Christine 380
Bealing, Paul 381
Bekker, Anriëtte (Annie) 31, 83, 380, 384
bell, ship's 21–2, 373–4
Belter, Jakob 384
Bengu, Captain Knowledge
  2019 Weddell Sea Expedition
    in 'Argentine' waters 86
    decision to end expedition 172–3, 176, 179, 182
    inspection of ship 66
    Master of Agulhas II 12, 379
    navigating through/around ice 22, 31, 54, 91, 132, 140, 170–1, 184
    stuck at King George Island 104
    weather concerns 194
  2022 Endurance22 Expedition
    anomaly on seabed 267
    captain's meetings 209, 234, 254, 260, 288, 308
    Endurance found 317–18, 319
    Master of Agulhas II 209, 383
    moon pool problem 251, 252
    navigating through/around ice 252, 254, 271, 285–6, 302
    at Shackleton's grave 369–70
    team members with Covid 225
    time extension 288
    weather concerns 285–6, 290, 291
Benham, Toby 32, 117–18
Bezuidenhout, Buzz 385
Big Spread, the 175–6
Biggs, Vincent 5

Binnie, Edward 349
bioluminescence 15
birthdays 105–6, 184–6, 253
Birtwistle, Nick 386
Blackborow, Perce Plate II.8, 135, 238, 299, 378
Blake, James 386
bongo nets 24–5
Bonin, Chad
  2019 Weddell Sea Expedition 49, 148, 154, 159, 161, 201, 204, 381
  2022 Endurance22 Expedition
    in Cape Town 207, 212, 374–5
    difficulties/problems 258, 259, 270–1, 272, 274
    Endurance found 310–11, 311–12, 315, 316, 320, 322–3
    in the ice pack 254
    live broadcast to school 230–1
    optimism 254, 290, 305
    position in team 384
    weirdness of Antarctica 305
Bornman, Tommy 380
Bound, Jo 73, 156, 215, 252, 275, 298, 322, 361
Bound, Mensun
  2014–15 Scharnhorst expedition 125, 201, 220–1
  2019 Weddell Sea Expedition see Weddell Sea Expedition (2019)
  2022 Endurance22 Expedition see Endurance22 Expedition (2022)
  birthday aboard Agulhas II 105
  Director of Exploration 379, 383
  eating penguin 60
  fate tied to mission 249–50
  Fellow of Explorers Club of New York 147
  idea for search for Endurance 17–18
  maritime experiences 5, 19, 115, 190, 194, 195–6
  other wreck sites 28–9, 43–4

Outward Bound Tours 97–8
  photograph **Plate II.9**
  Sandefjord visit 73
  Shackleton, fascination with 5
Boundbooks bookshop, Falkland
  Islands 97
Bransfield Strait 78, 96, 98, 99, 107, 111
Brashar, Kevin 385
Brokensha, Grant 309, 386
Brophy, Jodi 385
Burden, Nicholas 386
Burger, Jessica 380
Burtt, Chad 385
Busche, Thomas 384

cabin condensing 301
Caillens, Jean-Christophe (JC)
  anomalies on seabed 265, 266, 267,
    271–2, 296–7
  Endurance found 311–18, 326–7
  photograph **Plate II.9**
  position in team 221, 234, 383
  Sabertooth checks/tests 243, 250
  Sabertooth problems 258, 268
Cape Town, South Africa 10, 12, 14,
    201, 204, 207, 232, 362, 374–5
Carmichael-Brown, Saunders 386
Carstens, Darius 385
Chase, Julianne 147
Cheetham, Alfred 247, 248, 342, 377
Christensen, Lars 72
Christie, Frazer 111–12, 132, 174–5, 380
chronometers 80, 120, 121–3, 255
Churchill, Winston 16, 224
circulation system, interoceanic 44–5
Clark, Grant 385
Clark, Robert S. 26, 113, 128, 185, 335,
    377
climate change 27–8, 29, 44–5, 71, 98,
    262
Coleridge, Samuel Taylor 139, 144
constipation 58

coring work 12, 36, 40, 42, 57, 183
Coulter, Dr Lucy 210, 234, 244, 268,
    300, 383
Countess of Ranfurly 77
Covid-19 210, 217, 225, 226, 234, 237,
    244, 268
Crean, Thomas 79–80, 247, 339, 343,
    377
Critchley, Mike 123–4
CTD (conductivity, temperature and
    depth) casts 24, 36, 40, 42, 57
Cummings, George C. **Plate II.15**
currents
  carrying icebergs/floes 34, 84, 97
  circumpolar 78, 79, 97, 242; *see also*
    Weddell Sea gyre

Darville, Ray 21, 30, 94, 142, 187, 381
Dattilo wreck, Tyrrhenian Sea 28–9
de Gerlache, Adrien 72
De Vos, Marc 228, 234–5, 290–1, 298,
    303, 308, 383
Delauze, Henri 221
depth of *Endurance* 43, 92
depth of Weddell Sea 112–13
depths of shipwrecks 43, 112
descendants of *Endurance* team 156,
    215, 240
Deutschland expedition (1912) 143
diaries, usefulness of 5–6, 23–4, 58, 61,
    280, 298, 340, 344, 345, 352
diatoms 99–100
Discovery expedition (1901–1904) 108,
    191
Dlamini, Kobla 252
dogs **Plate I.8**, 61, 64–6, 67–8, 128, 343
DOS (Deep Ocean Search) 221
Dowdeswell, Evelyn 380
Dowdeswell, Julian 379
Drift & Noise 218–19
drones 36, 50, 71, 168, 305
Dump Camp 152, 351–2, 352–4

Eclipse ROV 40, 54–5, 93–5
education 213, 230–1
Elephant Island ix, 61, 62, 78–9, 105–6,
    128, 129, 134–5, 136, 137–8, 299,
    324
Elkington, Carl 234, 237, 255, 256, 383
Endurance; see also Imperial Trans-
    Antarctic Expedition (1914–17)
    abandoned 32, 152, 176–7, 340, 352
    archaeo-biogeochemical
        importance 113–14
    binnacle Plate I.5, 343
    Bound's 2018 predictions 325–6
    bow Plate II.13, 91, 315, 331, 347
    Chippy's hut 335–6
    chronometers 80, 121–3, 255
    compasses 79, 333, 343
    condition on seabed 317, 325–6, 359
    construction 72
    counter-stern 331
    crushed by ice Plate I.7, Plate I.8,
        9, Plate I.9, 32, 280, 331–2, 336,
        337, 340–1, 344
    deckhouse 321, 331, 342, 344–5
    depth on seabed 43, 92
    diagram 329, 330, 331
    drift after sinking 119–21, 281, 298,
        303, 305–6, 308
    in dry dock Plate II.11
    flare gun 373
    fo'c'sle deck 331, 347
    forehatch 346
    funnel 152–3, 337
    galley and pantry 321, 331, 342
    gramophone 223, 248, 306
    historical importance 19–20
    icebound Plate I.7, Plate I.8, 9,
        Plate I.9, Plate I.14, Plate I.15,
        Plate I.16, Plate I.19, 169, 170,
        215, 335–6
    icebreaking 91–2
    Kelvin sounding machine 334–5

kennels 64, 141, 343
main deck 329, 331, 338, 341–6
main deck, holes cut in the 339,
    351–7
map of last voyage ix
masts and spars 41, 274, 280, 317,
    321, 325, 326, 337, 340, 341
model, Spence's 114
name and Polaris star on stern
    Plate II.12, 331, 359
pens for pigs/penguins 62, 341
photographs
    Hurley's compared to current
        335–6
    Hurley's record 19
    icebound 5, 22, 41, 280, 337
    under sail 339–40
    wreck on seabed 328–9, 370
poop deck 329, 332, 336–7, 340
poop deck cabins 338–41
protocols to follow if found 156
refitting for Antarctic 72
rudder 331–2, 333, 335
under sail Plate I.1, Plate I.5
sallying ship 141
salvage missions 336, 345, 346,
    351–7
satellite deposit 265–9, 271–2
scientists' cabin ('Rookery') 338–9,
    340
search for see Endurance22
    Expedition (2022); Weddell Sea
    Expedition (2019)
Shackleton's cabin 336, 339–40, 371
Shackleton's purchase and
    renaming of 72
ship's bell 21–2, 373–4
ship's wheel Plate II.14, 332–4, 360
sinking 9, 41, 119–20, 124, 152–3, 337
skylight 336–7
stern Plate II.12, Plate II.14, 153,
    331–2, 335

stoves 340, 343, 355
strength 43, 325–6
tableware 344
wardroom ('Stables') 321, 331, 338, 342–3, 344
well deck 329, 332–6, 338, 360
wheelhouse 335–6
Worsley's coordinates 75, 118, 119–21, 123, 124, 125, 255, 263, 281
wreck on seabed **Plate II.6, Plate II.12, Plate II.13, Plate II.14**
Endurance painting (Cummings) **Plate II.15**
Endurance: The Hunt for Shackleton's Ice Ship documentary 241
Endurance22 Expedition (2022); *see also Agulhas II, S.A.*
    timeline:
        in Cape Town 207–13, 215–19
        voyage to Weddell Sea 223, 224–5, 226–7, 228–9, 230–1, 232–3, 234–5, 236–7, 240–1, 242–4
        entering Weddell Sea 245–6
        moving to search area 249–53, 254–6
        at search area 256–7, 258–9, 260–2, 263–4, 265–9, 270–3, 274–5, 276–7
        continuing search 278–9, 281, 282–3, 284–6, 287–9, 290–2, 293–5, 296–8, 300–3, 304–6
    Endurance found 307–23, 348–50
        going over predictions 324–6
        informing other team 326–8
        laser dive 328–9
        4K camera visual inspection **Plate II.6**, 329
            bow **Plate II.13**
            cabins below poop deck 338–41
            fo'c'sle deck 347
    main deck 341–6, 351–2
    poop deck 336–7
    stern **Plate II.12**, 331–2
    well deck **Plate II.14**, 332–6
    last dive 348
    celebration on the ice 350
    leaving the search area 350–1, 358–60, 361–3, 364
    arriving back at Grytviken 368, 369
    at Shackleton's grave 369–72
    sailing for Cape Town 372, 373
anomalies/POIs 213, 265–9, 271–3, 276–7, 279, 285, 287, 293–4, 296–7
back-deck team **Plate II.3, Plate II.10**, 220–2
Covid-19 210, 217, 225, 244, 268
critics' advice 283, 288–9, 297
Drift & Noise team 218–19
field teams 224
football on the ice 260, 284, 350
helicopters 210–11, 217–19, 243, 255–6, 350
ice buoys 255, 256, 262
ice projections 261
iceberg/floe threats 210, 270–1, 288–9, 291, 293
John and Mensun's evening meetings 235, 262, 272, 278–9
live broadcast to school 230–1
media 263, 283, 288, 358, 359–60, 361–2, 364
media team **Plate II.2**, 211, 224, 263, 386
meetings and presentations 209, 227, 228, 232–3, 234–5, 236–7, 241, 254, 260, 267–8, 274–5, 297, 308
mission 224
music 306, 324
100th anniversary of Shackleton's burial 348–50
reports to trustees 252, 278, 279

Endurance22 Expedition (2022) *cont.*
   Sabertooths *see* Sabertooth
     search vehicles
   satellite imagery, use of *see* satellite
     imagery, use of
   search box 255, 324–5
   team **Plate I.3**, 383–6
   time extension 283, 284, 288
   What would Shackleton think?
     291–2
   White Desert team 219, 237
Evans, Jeff 380
Ewart, Holly 57, 157, 381
Explorer 97–8
Explorers Club of New York 147

Falkland Islands 5, 39, 60, 97, 99, 362
Falkland Islands Maritime Heritage
     Trust (FMHT) 201, 220–1,
     230–1, 268, 327, 358, 359–60
fast ice 12
Fawcett, Sarah 108, 189, 379
February, Orlando 168, 252
Filchner, Wilhelm 143
First World War 9, 278–9, 368, 370
Fit for a FID cookbook 60
Flora Antarctica (Hooker) 99–100
Flotilla Foundation 118, 177
Flynn, Raquel 380
foodchains, Antarctic 100
football on the ice 14–15, **Plate I.19**,
     Plate I.20, 260, 284, 350
Fram 43, 332
Franklin, Sir John 146
frazil ice 16
FRC (fast rescue craft) 34, 36, 50
freezing temperatures of water 16
Frinault, Betina 381
frostbite 62, 299–300
Furious Fifties 190, 230, 242

gases, greenhouse 29, 45

Gilbert, Simon 123
Green, Charles J. **Plate I.17**, 127, 129,
     135, 365, 377
Greenstreet, Lionel 128, 185, 299, 377
Grogan, Dr Claire 92, 131–2, 157, 379
growlers 20, 196
Grytviken, South Georgia **Plate II.15**,
     20, 366, 368, 369
Guy, Emmanuel 385

Halstead, Chad 385
Heart of Antarctica, The (Shackleton)
     59, 116
Hebridean Sky 102
helicopters 210–11, 217–18, 219, 243,
     255–6, 350
Henry, Tahlia 379
Hewit, Natalie 211, 234, 369–70, 383
HiPAP (high-precision acoustic posi-
     tioning) calibration 34, 53, 157,
     159, 162, 163, 171, 174, 180
HiSAS (high-resolution interfero-
     metric synthetic aperture sonar
     survey) 151–2
Holness, Ernest 109, 110, 144, 378
Hooker, Joseph Dalton 99–100
Hope Point, South Georgia 367
Horvath, Esther 386
How, Walter Ernest 22, 377
Howard, Blake 148, 161, 381
Hudson, Hubert T. 11–12, 192, 377
Hughes, Timothy 385
Hurley, James Francis (Frank)
   disputes and squabbling 127, 134–5,
     138
   dog-training 67
   films and photographs 3, 5, 19, 41,
     46, 89–90, 280, 335, 337, 339–40,
     341
   ice-damage to *Endurance* 341
   killing/eating penguins 62
   last official visit to *Endurance* 373

music 248, 334
photograph of **Plate II.7**
position in team 89, 377
salvaging food from *Endurance* 353, 354, 355
seals 46
stage for theatricals 191
survival 116
Huseyin, Tim 381
Hussey, Leonard D. A.
anemometer 335–6
meteorological log 305–6
music 105, 246–7, 248, 342, 357
position in team 377
Quest expedition (1921) 365
Shackleton's death 366, 367
Shackleton's funeral 349
at the wheel 333
Hutchinson, Katherine 44–5, 379

ice loss 27, 45, 71
ice, types of 12, 16–17, 24
icebergs and floes; *see also* pack ice
A-68 31–2, 35–6, 36–7, 66, 210
A-69 210
giant floes 172–3, 174–5, 270–1, 291, 293, 294, 296
greybeards 243
growlers 20, 196
Kinkering Congs 199
photographs **Plate I.11**, **Plate I.14**
tabulars 15, 17, 83–4, 199, 210
icebreaking techniques **Plate II.1**, 91–2
'If' (Kipling) 340, 371
Imperial Trans-Antarctic Expedition (1914–17); *see also Endurance*; *specific members*
timeline:
aboard *Endurance*
birthdays 105, 184–5
Churchill's order to

proceed 16
disputes and squabbling 136
dogs 343
first encounter with ice 243
football on the ice 14
harpooning blue whale 239–40
icebreaking 91–2
James's metalwork 114
meetings, lack of 236
music 246–8, 306, 342–3
New Year's celebrations 11–12
sailing from South Georgia 368
trapped in Weddell Sea ice 184–5
voyage from England to South America 223–4
aboard icebound *Endurance*
abandoning ship 32, 152, 176–7
amateur theatricals 191–2
Chippy's hut 335–6
dogs moved to ice floe 64–6
King's birthday 105
marine biology work 113
occulations 122–3
penguins, killing and eating **Plate I.17**
sampling, biological and geological 335
seals, filming 46
seals, killing and eating 240, 247
camping on the ice
Chippy's hut 336
Chippy's insubordination 109, 236
constipation 58
despair 134
dict 58

Imperial Trans-Antarctic Expedition (1914–17), camping on the ice cont.
    disputes and squabbling 128–9, 136–7
    Dump Camp 152, 351–2, 352–4
    food situation 352–3, 356–7
    frostbite 299
    ice floe breaking up 110, 144, 309
    icebergs 84
    last official visit to Endurance 373
    latrines 58
    music 246–8
    Ocean Camp 119–20, 124, 136–7, 152, 280, 336, 343, 345, 351, 352
    Patience Camp **Plate II.7**, 67
    Paulet Island, attempt to reach 86–7
    penguins, killing and eating 185
    salvage missions 345, 351–2, 353–7
    Shackleton's birthday 185–6
    shooting Mrs Chippy 33, 109, 299
    shooting the dogs 67, 128
    sinking of Endurance 9, 152–3
    sunsets 84
    throwing away possessions 32
    treks 32–3, 352
lifeboat journey to Elephant Island 77–9, 299
on Elephant Island
    birthdays 105–6
    disputes and squabbling 128, 129, 134–5, 136, 137–8
    frostbite 299
    hunger 134–5

penguins, killing and eating 61, 62
thefts 138
James Caird journey to South Georgia 79–82, 109, 110, 192, 241, 247
rescues 324
diaries 5–6, 23–4, 35, 58
dogs 64–6, 67–8
First World War's impact 278–9
football on the ice **Plate I.19**, 260
map of ordeal ix
objective 3, 9
overview of ordeal 3–4
photographs **Plate I.1**, **Plate I.5**, **Plate I.6**, **Plate I.7**, **Plate I.8**, **Plate I.9**, **Plate I.14**, **Plate I.15**, **Plate I.16**, **Plate I.17**, **Plate I.19**
planning and preparation 3, 33
Polar Medals 108–10
questions about purpose 224
Shackleton's account in South 23, 32–3, 58–9, 61, 68, 116, 239–40
social stratification 127
survival 4, 9–10, 115–16, 351–2, 375
team **Plate I.1**, 377–8

Jacob, Tim 213, 215, 386
Jacobsen, Frithjof 366
Jakobsen, Johan 72
James Caird boat 79–82, 109, 110, 192, 240–1, 247, 368
James Clark Ross, RRS 31
James, Devon 148, 161, 381
James, Reginald W. (Jimmy)
    beauty of ice 304
    diary 23
    disputes and squabbling 128, 137
    Endurance's objective 115
    food situation on ice 352
    ice floe breaking up 144
    interview for job 246

metalwork 114
navigation 122, 124–5
pen freezing/bursting 300
position in team 377
salvaging food from *Endurance* 354–5
Johnson, Zakaria 385

Katlein, Christian 308, 384
Kelvin and James Limited 334–5
Kelvin sounding machine 334–5
Kerr, Alexander J. 136, 365, 377
King George Island 94–5, 98, 101
King, Olive 382
Kingsford, John 221
Kongsberg of Norway 40, 47, 149
krill 27, 99, 100, 242

La Harpe, Colin de 381
Lamont, Donald 156, 220, 272–3, 311, 348, 359, 362
Larsen B ice shelf 71
Larsen C ice shelf ix, 10, 22, 31–2, 34, 35, 63, 68–71
Larsen D ice shelf 210
Larsen Inlet 73–4, 83
Larsen, Ole Aanderud 332
Last Man Off (Lewis) 196
latitude, calculating 80–2, 120–1, 123–5
Le Gall, Pierre 161, 272, 277, 310, 312–18, 373, 381, 385
Leek, Joe 266, 268, 326, 384
Ligthelm, Captain Freddie
   2019 Weddell Sea Expedition
      arrival at search box 145–6
      decision to end expedition 176, 182
      ice captain of *Agulhas II* 12, 379
      monitoring ice conditions 34, 36, 66, 87, 90, 102, 117, 126
      navigating through/around ice 36–7, 130–1, 167–8, 169, 184, 197

penguins 130–1
   trapped in pack ice 140
2022 Endurance22 Expedition
   Endurance found 313–14, 317, 318, 319–20
   ice captain of *Agulhas II* 209, 383
   monitoring ice conditions 209–10, 234–5, 236–7, 270, 291
   navigating through/around ice 256, 261–2, 271, 302–3, 351
   sign with signatures and coordinates 350
Lindblad Explorer 97–8
Lindblad, Lars-Eric 97
longitude, calculating 81, 120, 121–2, 255
Loubser, Kobus 382
'Love Me Do' (Lennon/McCartney) 246
Lu, Liangliang 102, 130, 380
Lundberg, Lars 310–11, 385
Luyt, Harry 108, 379

McCarthy, Timothy 79–80, 82, 378
Macé, François (Fanche) 266, 310, 312–13, 315, 385
McGunnigle, Robbie 266, 310–18, 320, 384
McIlroy, Dr James A. 192, 365, 366, 377
Macklin, Dr Alexander H.
   abandoning ship 177
   birthdays 105
   diary 23–4, 35
   disputes and squabbling 89, 128, 129, 135, 136
   dogs 67–8
   essential articles 299
   forehatch on day of sinking 346
   frostbite monitoring 299
   grandchildren 323

Macklin, Dr Alexander H. *cont.*
 ice after sinking of *Endurance* 280
 ice-damage to *Endurance* 340–1
 personal possessions, loss of 341
 Polar Medal snubs 109
 position in team 377
 Quest expedition (1921) 365
 Shackleton's death 366–7
McLeod, Thomas 61, 365, 378
McNish, Harry (Chippy)
 attempt to repair rudder 332
 diary 23
 disputes and squabbling 127, 135–6,
  342
 eating penguin 61
 hut 335–6
 insubordination 109, 236
 James Caird journey to South
  Georgia 79–80
 King's birthday 105
 music 248
 pig and penguin pen 62, 341
 Polar Medal snub 108–9, 136
 position in team 377
 salvaging food from *Endurance*
  353–4, 356
 wardroom cubicles 343
Magabutlane, Thapi 382
March, Steven 54, 55, 93, 381
Marston, George E. 247, 248, 340,
 343, 377
mate during 2019 expedition
 'Argentine' waters 86
 decision to end expedition 177, 200
 describing ice 90, 91, 126, 271
 icebound 140, 158
 Mensun's friendship with 12–13,
  37–8, 202
 navigating through/around ice 40,
  88, 170–1, 174, 184, 187, 199, 302
 storm 189, 195
Matthee, James-John 380, 384

Mawson, Douglas 89
media 92, 186, 211, 220, 224, 263, 269,
 283, 288, 358, 361–2, 364
microscopic life 99–100, 113
Montelli, Aleksandr (Sasha) 380
Montevideo, Uruguay 201, 223, 321,
 339, 367
Morgan, the chief steward 323
Morizet, Grégoire 266, 326–7, 328–9,
 385
Morizet, Jérémie (Jim) 266, 312, 313,
 315, 319, 385
Morrell, Benjamin 143
Morris, Paul 386
Mrs Chippy **Plate II.8**, 33, 109, 223, 299
Murashkin, Dmitrii 384
music 105, 127, 191–2, 223, 245, 246–8,
 306, 324, 334, 338, 342–3, 357, 365

Nansen, Fridtjof 100
'Narrative of Arthur Gordon Pym'
 (Poe) 142–3
National Maritime Museum,
 Greenwich 247
navigation techniques/tools 38, 76,
 78, 79–80, 80–1, 112, 123, 334; *see
 also* latitude, calculating; longi-
 tude, calculating; satellite
 imagery, use of
Nelson, Admiral Horatio, 1st Viscount
 19–20
New Island 60
Nimrod expedition (1907–1909) 59, 116,
 146, 179
Nordenskjöld, Otto 33, 86

occultations 122–3, 255
Ocean Camp 119–20, 124, 136–7, 152,
 280, 336, 343, 345, 351, 352
Ocean Infinity 47, 48, 57, 147
O'Hara, Dave 54, 55, 94, 381
Onde, Maeva 266, 268, 326, 327–8, 384

Orde-Lees, Thomas H.
anguish about War 278
birthday 105–6
bullying and humiliating 127, 129,
134, 135–8, 334
captain's cabin 339, 340
Chippy's hut 336
cold 284
constipation 58
construction at Ocean Camp 345
diary 23
disputes and squabbling 128, 342
eating penguin 62
Endurance crushed by ice 332, 337
forehatch accident 346
frostbite 300
iceberg threat 170
music 247, 248
parhelia and sun pillars 282
passageways 338
position in team 377
salvaging food from *Endurance* 352,
353–4, 357
seals 240, 247
Shackleton's birthday 185
ship's wheel incidents 333–4
shooting of the dogs 65
snoring 128, 136, 137–8
theatricals on *Endurance* 191–2
throwing away possessions 32
O'Reilly, Mark 252
'other worlds' 142–3
Ottesen, Dag 380
Outward Bound Tours 97–8
Oxner, Todd 148, 160–1, 381

pack ice
the Big Spread 175–6
breaking through **Plate II.1**, 91–2
dynamism 164–5, 209–10
loss of, through climate change
27, 262

photographs **Plate I.2**, **Plate I.4**,
Plate I.7, **Plate I.8**, **Plate I.9**,
Plate I.12, **Plate I.14**, **Plate I.15**,
Plate I.16, **Plate I.19**, **Plate I.20**
parhelia 282
Patience Camp **Plate II.7**, 67
Patz, Michael 306, 385
Paulet Island *ix*, 32–3, 86
Penguin Bukta 10, 12, **Plate I.18**,
198–9, 200
penguins
chanting 176
curiosity 14–15, **Plate I.20**, 62,
181–2, 250, 259, 306
farewell 350
killing and eating **Plate I.17**, 60–3,
185, 341
playing 'chicken' 130–1
population declines 27
smell of excrement 284
ubiquity 24, 157, 200, 202, 254, 295
petrels 24, 83, 242, 358
phytoplankton 25
pigeons, Cape 242
plankton 15, 25, 113, 242
Poe, Edgar Allan 142–3
Polar Medals 108–9
Polaris see Endurance
Powell, R. C. F. 279
'Prospice' (Browning) 370–1

Queen Maud Land, Antarctica 10, 17,
Plate I.18, 198
Quest expedition (1921) 20, 364–5, 367

Rabenstein, Dr Lasse 210, 218, 228,
234, 244, 261, 262, 298, 303, 308,
383
Rack, Wolfgang **Plate I.12**, 83, 380
Ramjukadh, Carla-Louise 383
Reach the World 213
refraction 124–5, 179

Rhapsody in Blue (Gershwin) 245

Rickinson, Lewis 192, 248, 377

'Rime of the Ancient Mariner, The'
     (Coleridge) 139, 144

Rinderknecht, Beat 384

Roaring Forties 190, 230

Robertson Island ix, 32–3, 71, 73–4

Ross, James Clark 9, 86, 99

Ross, Tom 385

Rothes, Lucy Noël Leslie, Countess
     129

ROVs (remote operated vehicles) 21,
     31, 33–4, 40, 42, 50, 51, 54–5, 57,
     63, 68, 93–5

Rowett, John Quiller 364–5, 367

Royal Geographical Society 215, 325

Saab 213

Sabertooth search vehicles
     checks and tests 243, 250–1, 256–7
     description of 212–13
     diagram **214**
     dives 263, 264, 278, 282, 287, 294–5,
          296–7, 298, 304
          anomaly on seabed 265, 266–72,
               277
          Endurance **Plate II.6**, **Plate
               II.12**, **Plate II.13**, **Plate II.14**,
               310–17, 320, 321–2, 326, 328–9,
               331–47, 348
          problems 257, 258, 277, 290,
               293–4
          scanning quality 258–9
     lowering/raising 218
     photographs **Plate II.3**, **Plate II.4**,
          Plate II.10
     team 221

Saint Amour, Steven 55, 57, 68, 94, 381

sallying ship 140–1, 167–8

Samuel, Claire 33, 48, 161, 171, 172,
     379

Sandefjord, Norway 72, 73, 239

satellite imagery, use of
     2019 Weddell Sea Expedition 22,
          37, 87, 90, 92, 102, 111–12, 117, 126,
          132, 170, 172, 174–5, 179, 190, 197
     2022 Endurance22 Expedition
          209–10, 216, 228, 234–5, 243–4,
          261, 288–9, 308

Saunders, Edward 59, 351–2

Schapman, Clément 266, 268, 310–18,
     385

Scharnhorst expedition (2019) 201, 220

Scilly naval disaster (1707) 121

Scott, Captain Robert Falcon 17, 108,
     116, 179, 180–1, 191, 199

Scott-Fawcett, Stephen 253, 349–50

Scott Polar Research Institute (SPRI),
     Cambridge 23, 32, 112, 117–18,
     122, 215, 305–6

Screaming Sixties 190–1, 242

seals
     around *Agulhas II* **Plate I.10**, 24,
          83, 132–3, 141–2, 157, 181, 223, 295
     around *Endurance* 26, 46
     killing and eating 58, 67, 105–6,
          240, 247, 355
     population declines 27

search boxes, calculating 117–19,
     120–1, 123, 255

sextants 38, 80, 123–5

Seymour Island 86

Shackleton, Frank 224

Shackleton, Lady Emily 223–4, 367, 371

Shackleton, Sir Ernest Henry
     birthdays 106, 184–6, 253
     Bound's fascination with 5
     burial 20, 348–50, 367
     Cape Horn voyage as teenager 76
     death 20, 349, 366–7
     descendants 156, 215
     Explorers Club of New York 147
     Falkland Islands expeditions 5
     grave **Plate II.16**, 369–72

Heart of Antarctica, The 59, 116

illness, 1921 365

Imperial Trans-Antarctic
    Expedition *see* Imperial Trans-
    Antarctic Expedition (1914–17)

life to the fullest 368–9

music 246

navigating through/around ice
    199

navigation skills 76

Nimrod expedition (1907–1909) 59,
    116, 179

no beelines 88

Nordenskjöld expedition relief
    supplies 86

Orde-Lees, relationship with 135

photographs **Plate I.1, Plate I.5,
    Plate I.7, Plate II.7**

poetry, love of 306, 369, 370–1

Polar Medal snubs 108–10

pushing the boundaries 371–2

Quest expedition (1921) 20, 364–6

sketch of Antarctica **Plate II.5**

sledge flag 146

South 23, 32–3, 58–9, 61, 68, 116,
    239–40

23 Psalm 355–6

Shackleton TV series (Sturridge) 20

Shawaal, the steward 217, 273, 300, 323

Shears, John
    2019 Weddell Sea Expedition
        arrival at search box 146
        AUV 7 going AWOL 167
        end of expedition 177, 179
        football on the ice 14
        iceberg-floe collision 51
        leadership role 379
        meetings 92
        monitoring ice conditions 66
        'monster floe' 172
        Polar Medal 108
        protocols review 156

retreat through ice 184

ROV disaster 94

sailing towards search box 132

seals 142

Shackleton's birthday 184

sledge flag 146

storm 190, 193–4, 197

stuck at King George Island 104

testing ice floe for safety 181

time extension 69, 154

2022 Endurance22 Expedition
    anomaly on seabed 266, 267, 272
    in Cape Town 209
    discussion about wars 278–9
    dive one with *Ellie* 257
    Endurance found on 100th
        anniversary of Shackleton's
        burial 348
    *Endurance: The Hunt for
        Shackleton's Ice Ship* 241
    evening tea with Mensun 235,
        262, 272, 278–9
    football on the ice 260, 350
    Hussey's meteorological log
        305–6
    ice floes for camps 255, 256
    informed about discovery
        318–21
    leadership role 220, 383
    meeting back-deck team 221–2
    meetings 234, 274, 308
    music 306
    openness 236
    photograph **Plate II.9**
    post-discovery interviews 359,
        362
    at Shackleton's grave 369, 370
    team members with Covid 225
    The Thing message 309
    time extension 283, 284
    time on the ice 284, 308–10,
        314, 317

shipworm (*teredo navalis*) 39

Shovell, Vice Admiral Sir Cloudesley 121

Silver Cloud 209

Sir Ernest H. Shackleton Appreciation Society 253, 349

Sir Ernest Shackleton (Smith) 5

sledge flags 146

Smith, Shantelle 380

SMP (SnowMicroPen) 232–3

Snow, Dan **Plate II.2**, 224, 241, 263, 267, 306, 359, 362, 370–1, 386

Snow Hill Island ix, 32–3, 83, 86

Soul, Fred 266, 268, 326, 327, 385

sounding machines 334–5

South film (Hurley) 46, 89

South Georgia ix, 3, 4, 5, 20, 81–2, 343–4, 346, 348, 349, 365–6, 367, 368–71

South (Shackleton) 23, 32–3, 58–9, 61, 68, 116, 239–40

Spee, Admiral Graf von 125, 201, 220

Spence, Kurt 114, 380

squid, giant 113

Stephenson, William 109, 110, 378

Steyn, Ben 384

Stocker, Alexandra (Ali) 293, 384

Strange, Espen 48, 149, 161, 381

Stubbs, Tamara 382

Sudur Havid disaster 196

Suhrhoff, Mira 384

sun pillars 282

sunrises and sunsets 84

Swanepoel, Michiel 234, 383

'sweethearts and wives' toast 342–3

Swiftsure, HMS 77

Tait, Charles 256, 273, 385

Taylor, Kerry 266, 268, 290, 326, 385

Taylor, Michelle 68, 381

Teich, Eduan 385

terns 24, 242

Terra Nova 17

theatricals, amateur 191–2

Thomas, Channing 48, 70, 148, 161, 164, 167, 171, 172, 180, 379

Thomson, William, 1st Baron 334–5

timekeeper, maritime 121

Titanic, RMS 35, 129

Trincali, Julien 191, 381

Tuhkuri, Jukka 384

Tutanekai, SS 76

23 Psalm 355–6

UAVs (unmanned aerial vehicles) 168

Ukraine, Russia's invasion of 278, 362, 368, 370

Van Zijl, Christof 380

Venter, Waldo 385

Victory, H.M.S. 19–20

Vincent, John 79, 109–10, 138, 247, 378

Vincent, Nico

  anomalies on seabed 266, 285

  captain's meetings 234, 254, 308

  challenges, dealing with 226–7, 258, 260, 286

  Endurance found 312–18, 319, 362–3

  experience 220–1

  Gardanne, meeting in 221–2

  moon pool doors 251–2

  photograph **Plate II.9**

  Plans A and B 218

  role 221, 383

  Sabertooths 213, 221

  search area 281, 288

  at Shackleton's grave 369

Vogt, Warren 385

Wairoa 75–6

wars 278–9, 368, 370

Watts, Patrick 295

waves, extreme 194–6

'Wearing o' the Green, The' 223, 247

weather bombs 164–5, 309
weather routing 190
Weddell Sea Expedition (2019); *see also*
    Agulhas II, S.A.; Bengu, Captain
    Knowledge; Ligthelm, Captain
    Freddie
  timeline:
    at Penguin Bukta 10, 11, 12–13,
      14–15
    sailing to Larsen C ice shelf 17,
      20, 22, 24–5, 26, 30
    at Operations Zone 31–2, 33–4,
      35–8, 40, 42, 44–5, 47–50, 51–2,
      53–6, 57, 63, 66, 68–9, 70–1,
      73–4, 83–4
    sailing towards *Endurance* 85,
      86, 87–8, 89, 90, 91, 92, 93–5
    sailing towards King George
      Island 95, 96, 97
    at King George Island 98,
      101–2, 103, 104, 107
    sailing back towards *Endurance*
      107–8, 111–12, 114, 117–19,
      123–5, 126, 129–31, 132–3,
      138–9, 139–43, 144
    at Worsley's coordinates 145–52,
      154–66, 167–9, 170–3, 174–8
    leaving the ice pack 179–82,
      183–4, 187–8
    heading back to Penguin Bukta
      189–92, 193–6, 197–9
    journey's end 200
  achievements 178, 201
  AUVs *see* AUVs (autonomous
    underwater vehicles)
  birthdays 105
  decision to end expedition 176–8
  drones, use of 36, 50, 71, 168
  football on the ice 14–15, **Plate I.20**
  justification for 19–20
  lecture on interoceanic circulation
    system 44–5
New Year's celebrations 11
objective 10
obstacles to reach *Endurance* 85
origins of idea 17–18
photographs **Plate I.2**, **Plate I.4**,
    Plate I.10, **Plate I.12**, **Plate I.13**,
    Plate I.20
Polar Medal for Shears 108
press reports 92
procedure briefing 92
protocols if *Endurance* found 156
remembering Shackleton's death
    20
ROVs 33–4, 40, 42, 45, 63, 68,
    93–5
satellite imagery, use of *see* satellite
    imagery, use of
scientific work 12, 24–5, 36, 40, 42,
    45, 57, 63, 68, 71, 83
scientists sent home 95, 96, 98
search box calculations 117–19,
    120–1, 123
team **Plate I.2**, 10, 379–82
time extensions 69, 154
Weddell Sea gyre 36, 87, 111, 144, 155,
    215, 228, 242–3
Weddell Sea, map of 1914–1917 events
    ix
Weddell Sea seabed 112–13
Whalers' Cemetery, South Georgia
    20, 349–50
whales 26–7, 71, 98, 225, 239, 240, 259,
    284–5, 358
whaling industry 26–7
White Desert team 219, 237, 300,
    385–6
Wild, Frank
  disputes and squabbling 137–8
  eating penguin 62
  last official visit to *Endurance* 373
  music 247, 248
  New Year, 1915 11–12

Wild, Frank *cont.*
  photographs **Plate I.9**, 337
  position in team 377
  Quest expedition (1921) 365, 367
  salvage missions 336, 346, 353
  shooting seal with arrow 240
  shooting the dogs 65, 67–8
Williams, Paul 382
Wolf's Fang, Antarctica 10, 12, 118,
    198, 202
Woodall, Lucy 204, 381
Woodville 349, 367
Wordie, James M.
  achievements 215
  anguish about War 278–9
  arrows 240
  attempts to reach open water
    184
  constipation 58
  descendants 215, 216
  diary 23
  disputes and squabbling 134–5
  drift map of *Endurance* 122, 215,
    216
  food situation on ice 352, 357
  frostbite 299
  helmsmanship 334
  ice-damage to *Endurance* 340, 344
  music 247, 324, 342–3
  position in team 377
  'Rookery' 338
  salvage missions 345, 352, 355,
    356–7
  sinking of *Endurance* 153, 281
Wordie, Pippa and Roderick 215, 216

Worsley, Frank
  alcohol, allowing 339
  captaincy of *Endurance* 77, 339, 377
  despair 134
  diary 23
  disputes and squabbling 128, 136–7
  Endurance, abandoning 340
  Endurance, ice damage to 321
  Endurance, sinking of 152–3
  Endurance's last coordinates 75,
    118, 119–21, 122, 123, 124, 125,
    255, 263
  experience, pre-*Endurance* 75–7
  food situation on ice 357
  foreign master's certificate 76–7
  ice after sinking of *Endurance* 280
  music 247
  navigation skills 4, 76–82
  navigation tools 79–80
  New Year, 1915 11–12
  New Zealand Prime Minister's
    praise of 77
  photograph **Plate I.6**
  Quest expedition (1921) 365–6, 367
  salvage missions 345
  Shackleton's death 366
  stay on Falklands 5
  'Wreck of the Hesperus, The'
    (Longfellow) 296
Wuis, Leon 381

Yelcho tug 324

Zavodovski Island 239, 243
zooplankton 242